THE U.S. AIRCRAFT CARRIER INDUSTRIAL BASE

FORCE STRUCTURE, COST, SCHEDULE, AND TECHNOLOGY ISSUES FOR CVN 77

John Birkler
Michael Mattock
John Schank
Giles Smith
Fred Timson
James Chiesa
Bruce Woodyard
Malcolm MacKinnon
Denis Rushworth

Prepared for the
United States Navy ✦ Office of the Secretary of Defense

National Defense Research Institute

RAND

The research described in this report was sponsored by the United States Navy and the Office of the Secretary of Defense (OSD), under RAND's National Defense Research Institute, a federally funded research and development center supported by the OSD, the Joint Staff, the unified commands, and the defense agencies, Contract DASW01-95-C-0059.

Library of Congress Cataloging-in-Publication Data

The U.S. aircraft carrier industrial base : force structure, cost, schedule, and technology issues for CVN 77 / John Birkler ... [et al.].
p. cm.
"Prepared for the United States Navy and the Office of the Secretary of Defense by RAND's National Defense Research Institute."
"MR-948-NAVY/OSD."
ISBN 0-8330-2597-X
1. Nuclear aircraft carriers—United States—Design and construction. 2. Navy-yards and naval stations—Economic aspects—United States. 3. Shipbuilding—United States—Costs. 4. Shipbuilding industry—United States. I. Birkler, J. L., 1944– . II. United States. Navy. III. United States. Dept .of Defense. Office of the Secretary of Defense. IV. National Defense Research Institute (U.S.).
V874. 3.U83 1998
359.9 ' 4835—dc21 98-6126
CIP

Cover illustration by Christopher Bing

Published 1998 by RAND
1700 Main Street, P.O. Box 2138, Santa Monica, CA 90407-2138
1333 H St., N.W., Washington, D.C. 20005-4707
RAND URL: http://www.rand.org/
To order RAND documents or to obtain additional information, contact Distribution Services: Telephone: (310) 451-7002; Fax: (310) 451-6915; Internet: order@rand.org

PREFACE

This report documents the methods and findings of RAND research on the adequacy of the defense industrial base to support further construction of aircraft carriers and on the cost, schedule, and technology issues associated with building the next carrier, designated CVN 77. This research was funded by the Naval Sea Systems Command and was coordinated through the office of the Director, Acquisition Program Integration, in the Office of the Secretary of Defense.

The research was briefed before the Senate Armed Services Committee's Sea Power Subcommittee in April 1997, and the findings were cited in the Committee's version of the FY98 Defense Authorization Bill, whose provisions are consistent with the recommendations made here. The testimony is available as

> John Birkler, *Aircraft Carrier Industrial Base,* Santa Monica, Calif.: RAND, CT-142, 1997.

The research documented in this report was carried out within the Acquisition and Technology Policy Center of RAND's National Defense Research Institute, a federally funded research and development center sponsored by the Office of the Secretary of Defense, the Joint Staff, the unified commands, and the defense agencies.

CONTENTS

FIGURES

TABLES

SUMMARY

U.S. military strategy in the post–Cold War era calls for a fleet of 12 aircraft carriers. Because that is the number of carriers currently in operation and because carriers have a finite life, new ships must be built as old ones reach retirement age. Building a carrier requires the participation of thousands of firms and thousands of individuals. Scheduling production of the next carrier must take into account the availability of ship-construction funds, the work required simultaneously on other vessels, the service lives of operational carriers, and the potential for loss of important carrier-production skills over time. On the last point, the currently planned 2002 start of construction for the next aircraft carrier, designated CVN 77, will be seven years after the start of the previous one—relatively long by the standards of the last few decades.

Motivated by these issues, the Navy asked RAND to assess the problems the industrial base might encounter in producing the next carrier and to answer the following questions:

1. What constraints does the need to meet carrier force requirements place on start dates for CVN 77?
2. Of those start dates responsive to force requirements, which permits the most economical carrier construction? Do longer gaps between starts entail risks?
3. What are the implications of different start dates for the survival of vendors supplying carrier components to the shipyard? What are the implications for the cost of those components?
4. Are there any technologies or production processes not now employed in Navy shipbuilding that could permit significant savings in carrier life-cycle costs? If so, research and development will be required to adopt such technologies and processes to aircraft-carrier construction and operation. How much R&D do the potential savings justify?

Our key finding on the aircraft carrier industrial base is as follows: Newport News Shipbuilding—the United States' sole nuclear-powered carrier-construction facility—and the supporting industrial base throughout the United States are expected to retain the basic capabilities necessary to build large, nuclear-powered aircraft carriers into the foreseeable future, regardless of when or even whether CVN 77 is built. However, failure to start CVN 77 in the 2000–2002 time frame will inevitably lead to some decay in the quality of those capabilities and, hence, to increased costs, schedule durations, and risks when the next carrier is started. While other work may employ similar skills, the current and projected workload does not maintain the volume of skills to build CVN 77 if the ship is delayed and would require significant reconstitution costs for both shipbuilding skills and selected component suppliers. But the differences in costs and risks may not be as important as the implications for the carrier force structure.

SCHEDULE AND FORCE STRUCTURE

To maintain a constant force size, new carriers must be constructed to replace older ships as they retire (see Figure S.1). Our examination of CVN 77 production scheduling begins with an analysis of the relationship between the desired fleet size and the timing of CVN 77 construction. When fleet size is constant, the interval between carrier starts is determined by fleet size and retirement age. The planned retirement age for ships currently in the fleet is close to 50 years. To sustain a fleet size of 12, a new carrier would have to be completed every four years. But carriers have not been completed precisely at 4-year intervals. This means that, for a given build interval and fleet size, retirement ages of individual ships will vary. Figure S.2 shows how a 5-year build interval (versus a 4-year interval) would affect the retirement ages of ships now in the fleet. Specifically, waiting five years between ships will necessitate service lives for *Eisenhower* and *Vinson* that are longer than the 48-year age at which these vessels are expected to deplete their nuclear fuel. Refueling these carriers is a very expensive operation, so these dates may be taken as "drop-dead" dates for these ships.

The constraints that end-of-fuel dates impose on the timing of new-construction starts can be seen more clearly in Figure S.3. There, the black bars represent a sequence of construction periods beginning with the current plans for CVNs 76 and 77. The overlapping sequence continues in hypothetical 4-year intervals. For most ships, we assume the 6.5-year construction period currently planned for CVNs 76 and 77. However, we allow an extra year for the two succeeding ships, which will be the first two ships of a new class of carriers, designated CVX (CVXs 78 and 79). We give the ages of retirement for the ships

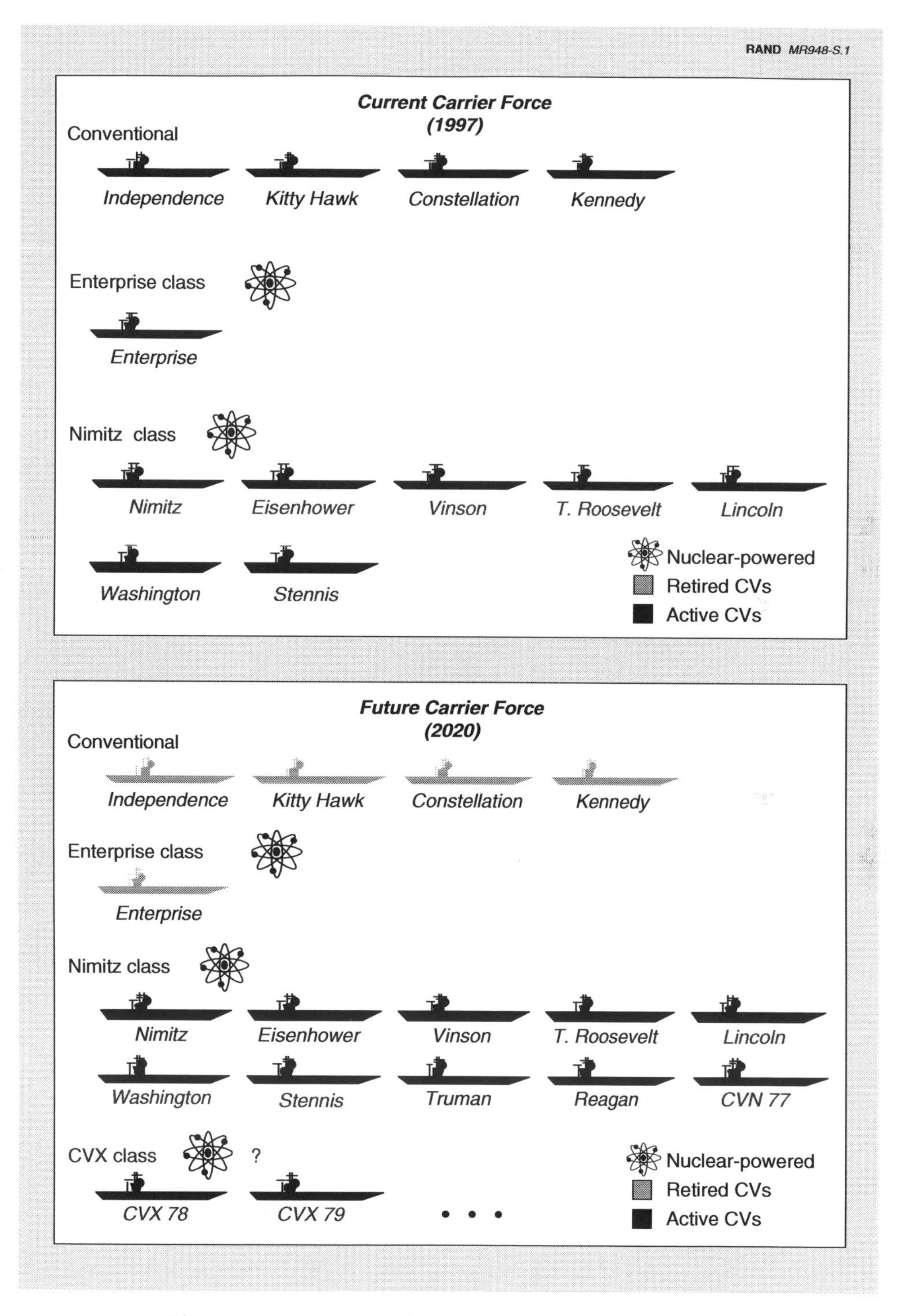

Figure S.1—Current and Future Carrier Force Structures

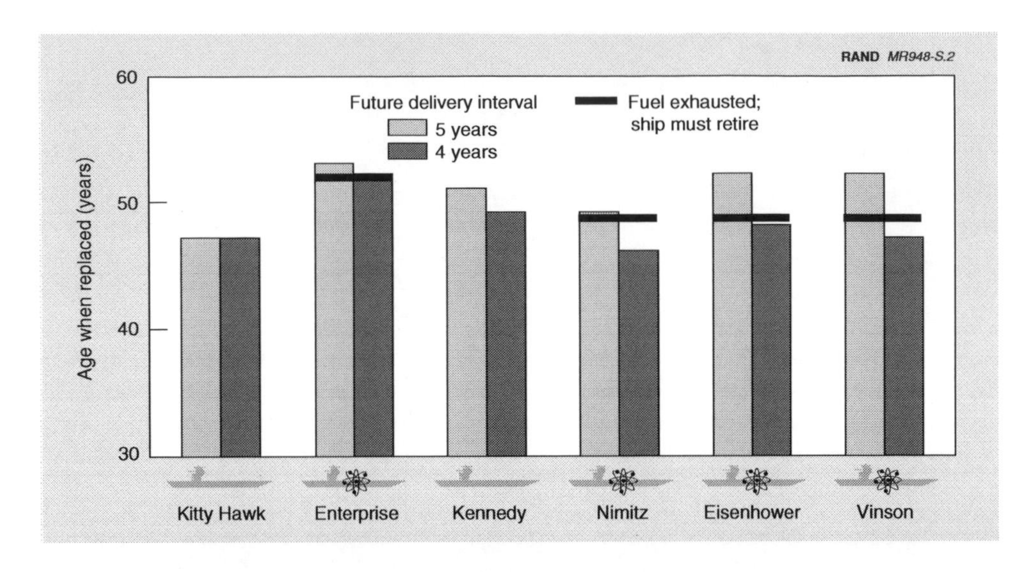

Figure S.2—Build Intervals for Sustaining a 12-Ship Fleet

being replaced, assuming retirement occurs as soon as a new ship is ready. We also show the dates (rather than ages) at which nuclear carriers are expected to run out of fuel.

As shown in the figure, starting CVN 77 in 2002 permits retirement of the ship it is replacing (CV 63) at age 47. CVN 77's start could thus be delayed a few years without placing the older ship at undue risk. However, there is only one Newport News Shipbuilding (NNS) dry dock in which carriers are constructed, which means that two carriers cannot be constructed simultaneously or nearly so, especially if one is just beginning the dry-dock construction phase and the other is nearing the end. Therefore, CVN 77's start date cannot be delayed beyond 2003 or 2004; otherwise, CVX 78's start date will also have to be delayed, and CVX 78 will not be finished before the ship it replaces, CVN 65, runs out of fuel. That end-of-fuel date and the one for CVN 69, in 2026, allow little leeway in start dates for the next several carriers if a 12-ship fleet is to be sustained.

SCHEDULE AND COST

Newport News Shipbuilding constructs nuclear-powered carriers and submarines, among other ships, and overhauls and refuels both carriers and submarines. The workload in the yard influences the costs of the ships it builds. If a major project comes to a complete end before another start, the yard incurs the costs of laying off workers, then rehiring and retraining workers, plus the inefficiencies associated with any new hires. If the second project begins as the

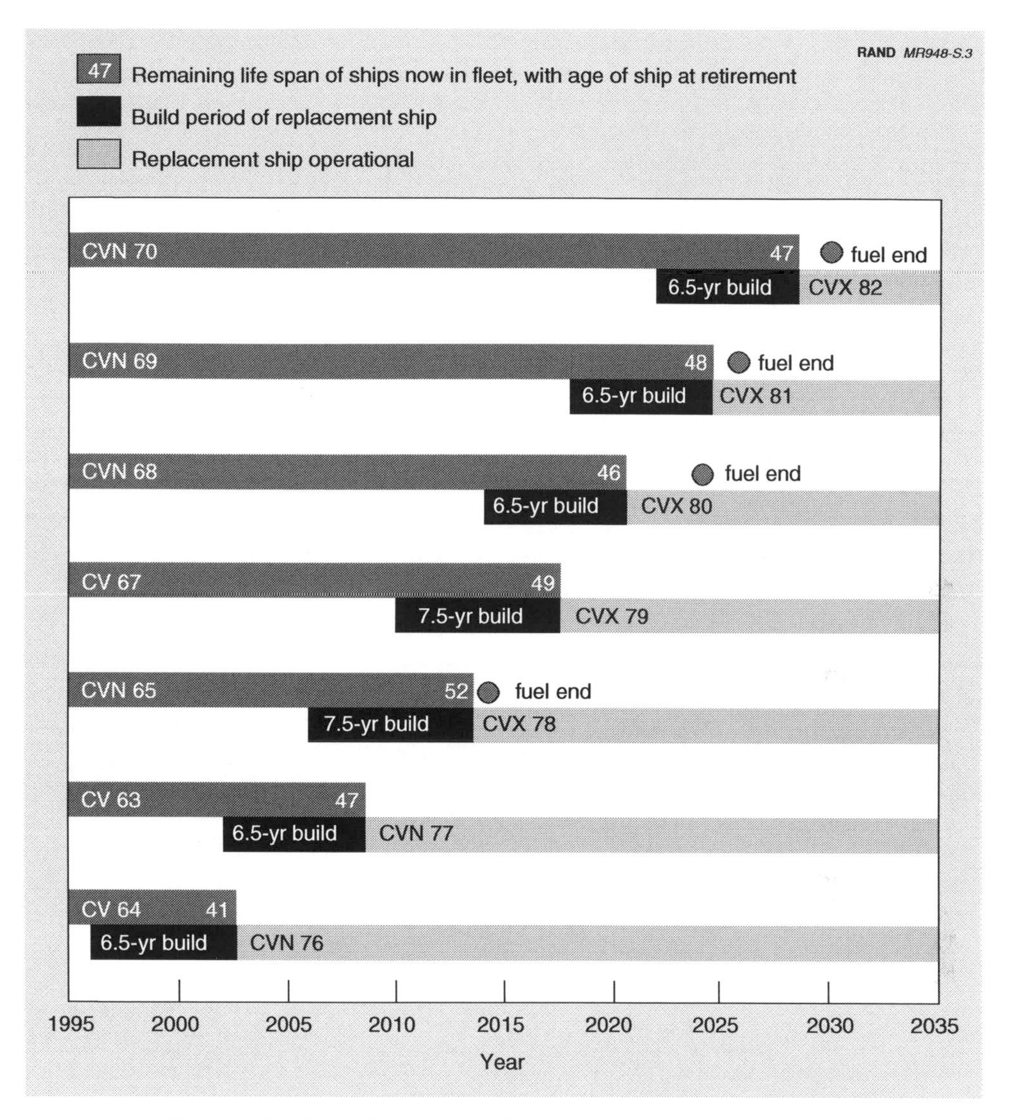

Figure S.3—End-of-Fuel Dates for Current Carriers in Relation to Nominal Ship-Construction Schedule

first is winding down, the yard can make an orderly transition of workers from the first project and the yard can avoid the costs associated with rebuilding the workforce.

We constructed a mathematical model that takes into account all the current and projected work at Newport News. The model calculates total shipyard costs as CVN 77's start date is varied against that fixed background of work.

Figure S.4 shows the cost implications of different build schedules for CVN 77.[1] (For completeness, we show costs for starts out to 2008, even though starts after 2004 cannot sustain the desired force structure.) Each point (diamond, square, or triangle) on the chart corresponds to the cost associated with a given build period and start date, minus the cost associated with the currently planned 6.5-year build starting in 2002. Thus, points above the zero line represent costs greater than those for the planned schedule; points below the line represent savings.

For any given build period, starting earlier within the window of interest (1999 through 2004) means lower costs for the yard. Labor level in the yard is the reason for those lower costs, as Figure S.5 illustrates. The demand profile for CVN 77[2] is shown to scale, with a start in 2002, over the labor-level profile for the other projects in the yard. The labor level from other projects dips between about 2002 and 2007; so, if CVN 77 is started before 2004, the additional demand flattens out the total shipyard workforce curve, decreasing costs associated with workforce swings.

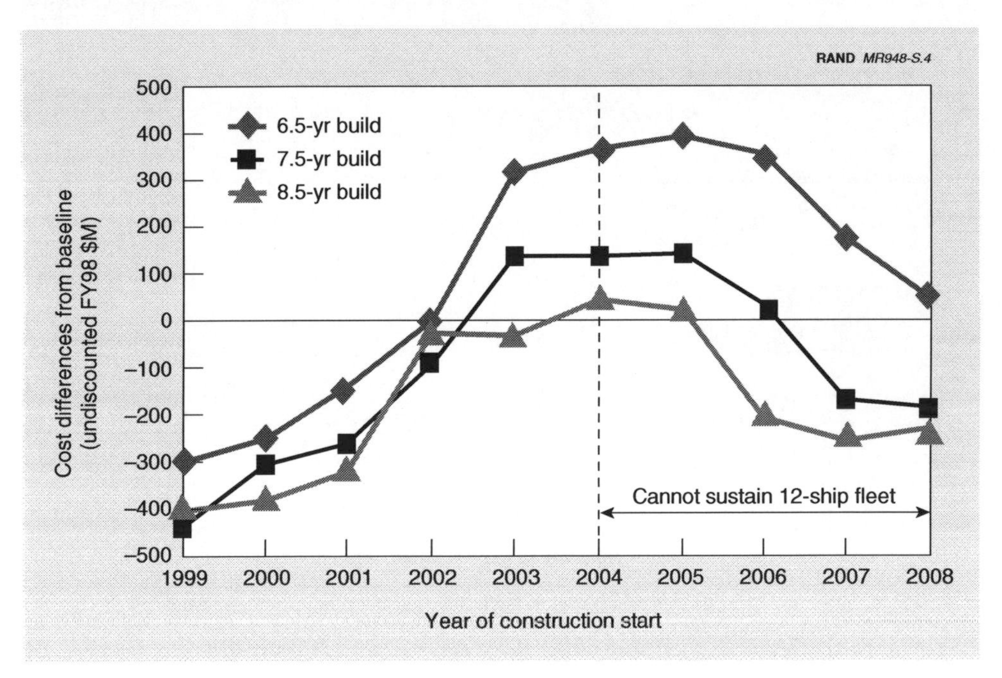

Figure S.4—Effect of CVN 77 Start Date and Build Period on Total Shipyard Costs

---

[1]All costs in this report are in FY98 dollars.

[2]Assumes a 6.5-year build period preceded by a 1-year engineering period employing few workers.

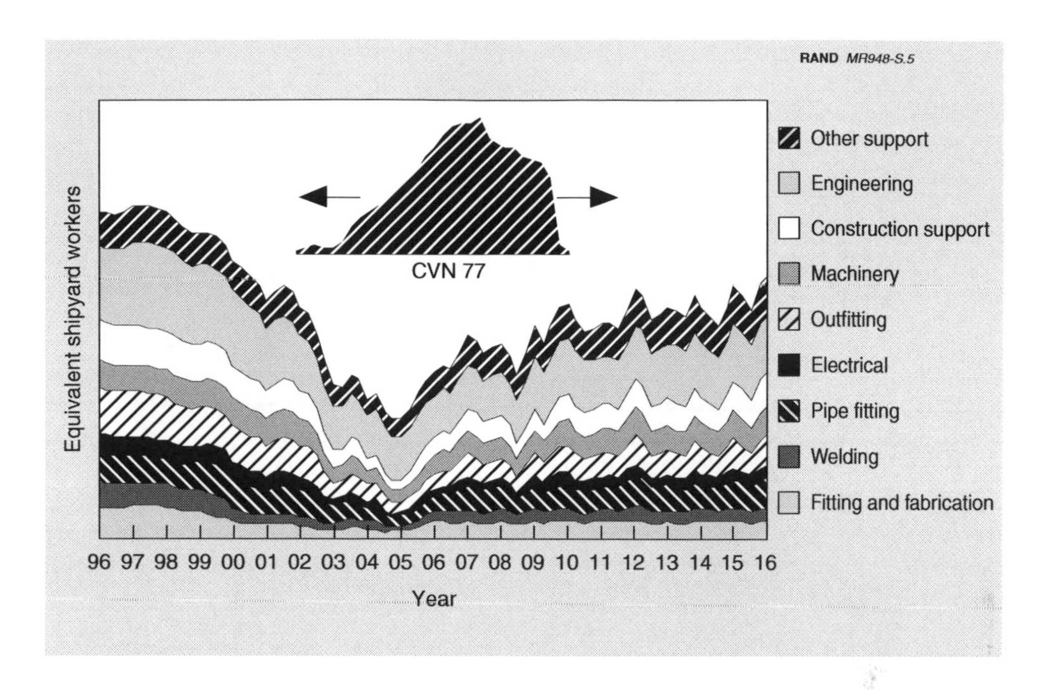

Figure S.5—The CVN 77 Labor Demand and the Total Shipyard Workforce Profile

It generally costs less if, for any given start date, CVN 77 construction is stretched over a longer period. Extra savings from longer build periods are due both to smoothing out of labor levels (see Figure S.6) and to other efficiencies associated with longer build periods—up to a point; 8.5 years is judged to be optimal. Note that, in moving from a 6.5-year build starting in 2002 to a 7.5-year build starting in 2001 and an 8.5-year build starting in 2000, costs decrease (see Figures S.6 and S.4). That is, if construction is to end in 2008 as currently planned, it costs less to start earlier than 2002.

SCHEDULE AND VENDOR EFFECTS

The shrinking of the commercial nuclear-power market, together with a reduction in Navy orders, has led to a drastic downsizing of the vendor base for the U.S. nuclear-power industry, including the base for Navy nuclear equipment—to as little as a single supplier for each major type of naval nuclear equipment. But manufacturers of light equipment and reactor cores for the Navy will remain viable at their present scale by meeting expected demands

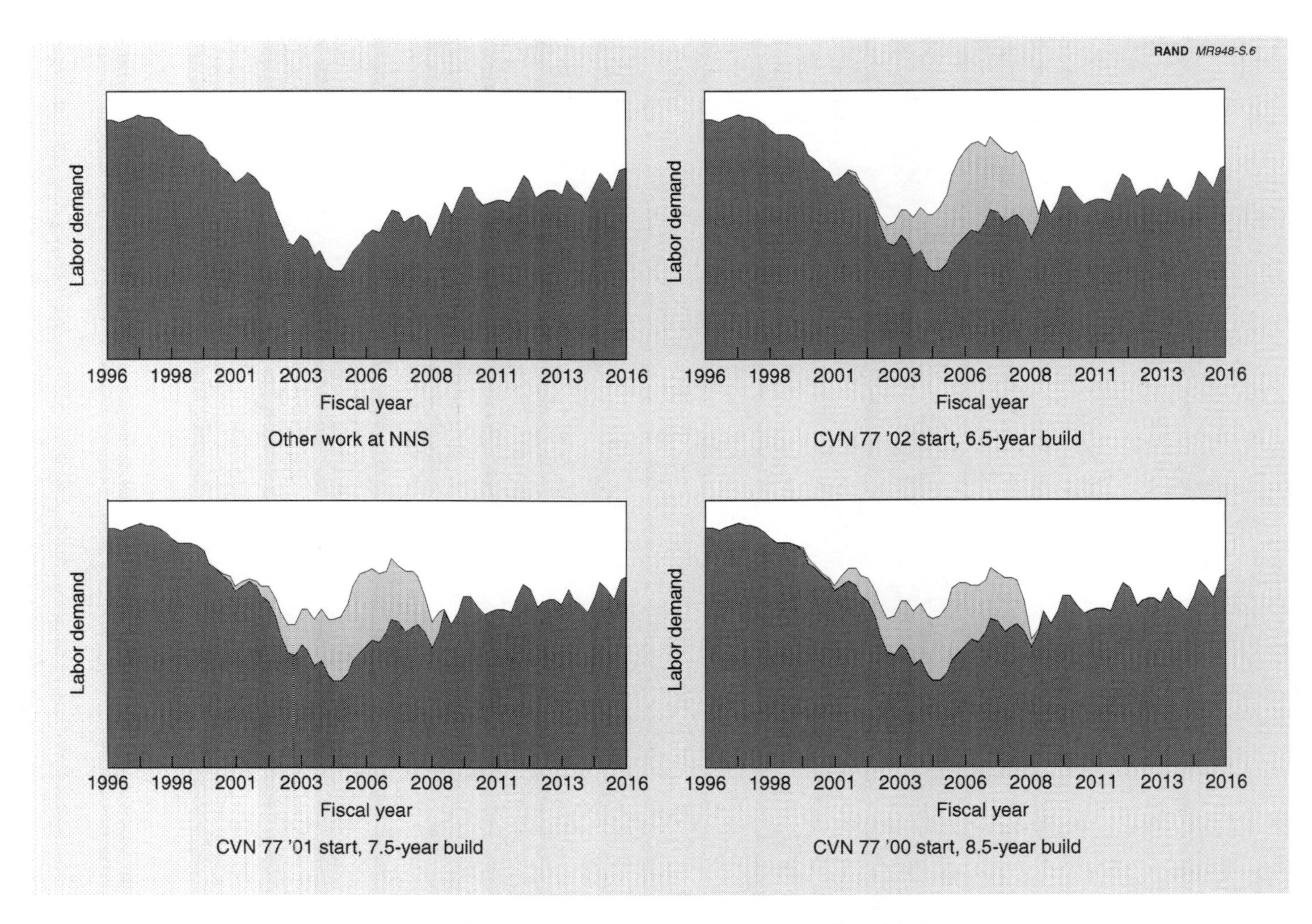

Figure S.6—Labor-Demand Requirements for Construction of CVN 77 for Different Start Dates and Build Periods, in Relation to Other Work at NNS

unrelated to carrier construction (e.g., new attack submarine and carrier refuelings).[3]

The same holds true for the sole Navy nuclear-related heavy-equipment manufacturer, which has already built the major reactor-plant components for CVN 77. Early in the construction of the current (Nimitz) class of carriers, an extra set of major reactor-plant components was funded as backup. Every time a carrier has been built, the backup set has been used in the new ship and the reactor-plant funding has permitted the construction of a new backup. CVN 77 is the last of its class; therefore, some of the spare set built during construction of CVN 76 will be used. Only a partial replacement spare shipset of reactor-plant components is planned for procurement with CVN 77 funds. Thus, when or even whether CVN 77 is built will not affect the heavy-equipment vendor in the near term.

The same basically holds true for nonnuclear vendors. No major product lines are in jeopardy. But some firms are having trouble sustaining a workforce adequate to supply carrier components through the current long gap between CVNs 76 and 77. The longer that gap persists, the more it will cost to reconstitute these capabilities. Therefore, orders for major long-lead-time items for CVN 77 should be placed no later than 2000, and possibly as early as 1998. Ordering early achieves modest savings—$50 to $80 million—by obviating the need for retraining the vendor workforce or taking other reconstitution measures.

SAVINGS FROM NEW PRODUCTION PROCESSES AND TECHNOLOGIES

We interviewed commercial shipbuilders to determine whether they were employing new technologies or production processes that might be applied in Navy-funded shipbuilding. The U.S. commercial shipbuilding sector has all but vanished. Necessarily, then, our interviews focused on foreign shipbuilders, and specifically on builders of cruise ships, because these are the closest analog to carriers in supporting large numbers of people for extended periods.

We did not find any technological advances that would result, *individually,* in large cost savings if implemented by U.S. shipbuilders. But foreign commercial shipbuilders are taking advantage of a wide range of existing production processes and technologies that have not been implemented in shipbuilding funded by the U.S. Navy. These shipbuilders appear to be saving substantial

[3]The light equipment and reactor cores are replaced during the midlife reactor refueling and overhaul. The Nimitz-class carriers are just starting to reach the midlife refueling point, thus creating a demand for light-equipment components and reactor cores.

amounts of money through large-scale outsourcing, especially for ship parts that can be supplied and installed as modules, and by close coordination with contractors for just-in-time delivery, among other things. By reducing the need for maintenance or the number of personnel required on board, some of the approaches taken save money not just during construction but also over the life of the ship.

Most of these production processes and technologies will require some research and development on the Navy's part if they are to be adapted for warships such as carriers. However, investing in R&D now could result in cost savings later, once CVN 77 is operational. We sought to determine how much of an R&D investment could be justified for improvements that would reduce two specific types of future costs:[4] costs associated with major maintenance activities scheduled to occur as the ship periodically becomes available in the shipyard, and costs associated with the ship's enlisted crew. We calculated costs not only for CVN 77 but for the remaining lives of other ships in its class, because some technologies adapted for CVN 77 might be backfitted to earlier Nimitz-class ships. Total costs are displayed by year in Figure S.7.

We calculated the present value of major shipyard availabilities as $38 billion and enlisted-crew costs as $27 billion, discounted to account for the lower present value of future dollars. Even if all possible technologies could save only 10 percent in operating costs, we estimated a potential savings of over a quarter billion dollars from CVN 77 alone (see Figure S.8) and about a half billion dollars from the entire class (after taking a large deduction for nonbackfittable improvements).

RECOMMENDATIONS

The analyses described above support the following recommendations:

- Begin CVN 77 fabrication before 2002. The potential for savings here is substantial—in the hundreds of millions of dollars.
- Order some contractor-furnished equipment in advance of shipyard start. This should permit additional savings in the tens of millions of dollars.
- Invest at least a quarter billion dollars in research and development directed at adapting production processes and application-engineering improvements that could reduce the cost of carrier construction, operation

[4]Note that, because our cost estimates are limited to these two categories, they are conservative; so, then, are our estimates of the R&D investments justified by future costs.

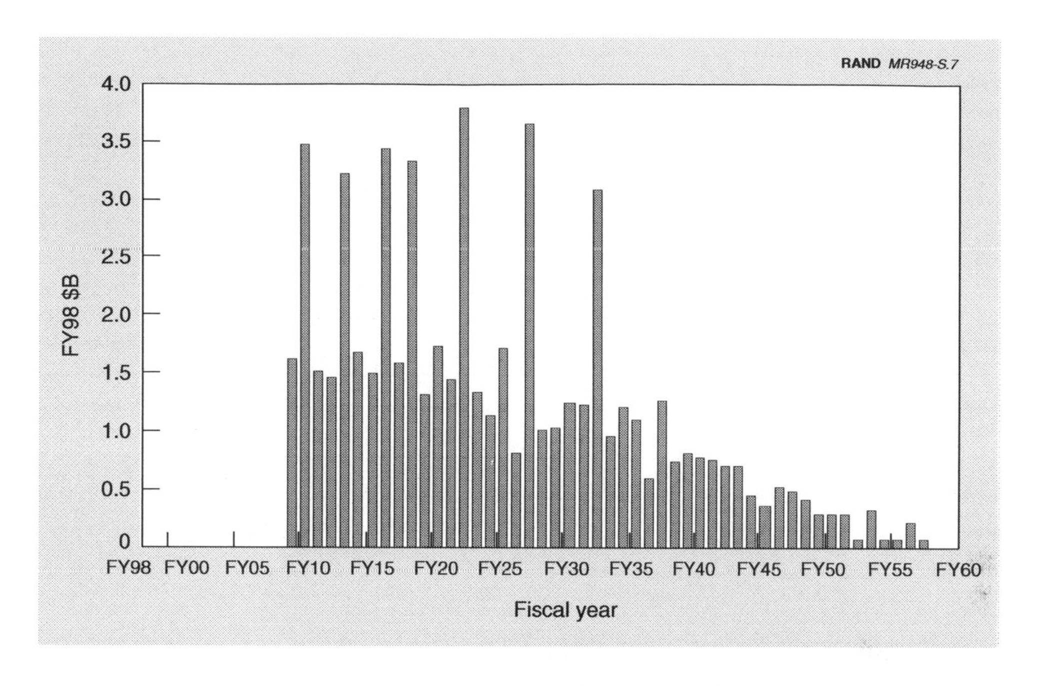

Figure S.7—Annual Costs for Scheduled Availabilities and Ship's Enlisted Company, Nimitz Class

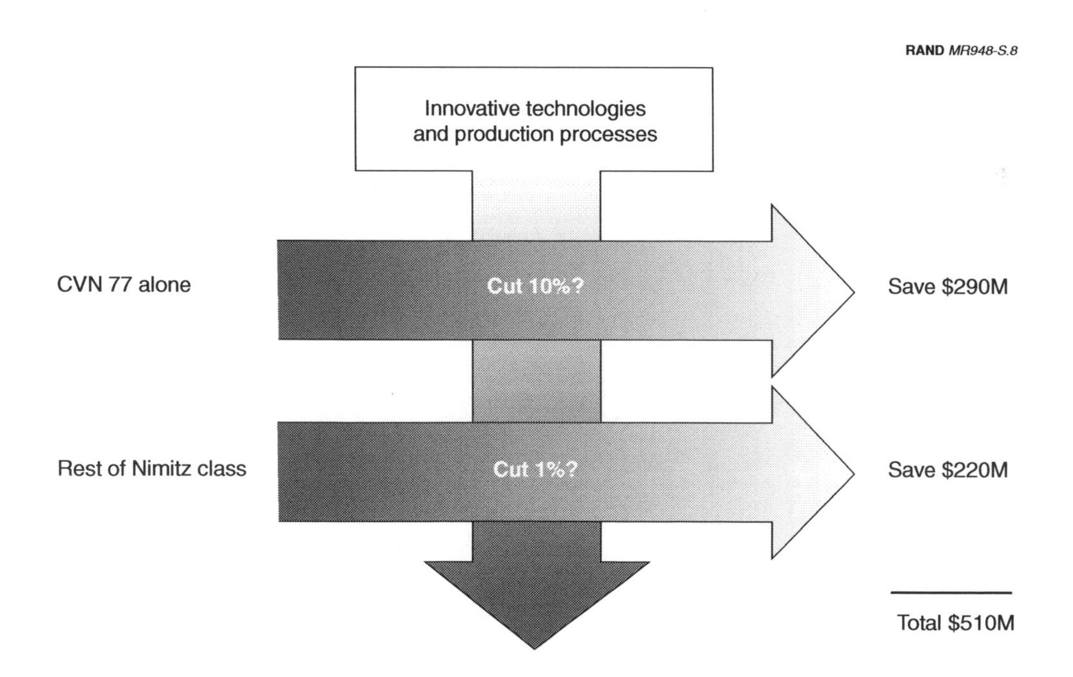

Figure S.8—Initiatives to Reduce Maintenance and Crew Costs

and maintenance, and manning. In fact, the costs involved in building and operating carriers are so large that the Navy should consider establishing a stable annual R&D funding level for these ships.

ACKNOWLEDGMENTS

This work could not have been undertaken without the special environment at RAND, which has been created and nurtured by RAND leadership. For that environment, the authors are grateful. Many individuals in the Navy and at RAND deserve credit for the work discussed in this report. Their names and contributions would fill several pages. If we were to single out a senior person in the Navy and another at RAND who participated in and supported this work in extraordinary ways, we would mention CAPT Mark O'Hare, USN, Program Manager for CVs/CVNs (carriers and nuclear carriers) at the Naval Sea Systems Command, and Gene Gritton, Acting Director, NDRI.

We also want to thank the leaders and staff of the Offices of the Secretary of Defense and Secretary of the Navy; the Naval Sea Systems Command—in particular, the Nuclear Power Directorate; Program Executive Officer (PEO), Aircraft Carriers; Naval Sea Systems Command Detachment PERA (CV); Puget Sound Naval Shipyard; and Norfolk Naval Shipyard. And last but not least, we thank the managers, engineers, and analysts at Newport News Shipbuilding. We extend our gratitude to each for arranging for us to visit their facilities and discuss production issues, for sharing experiences of those most directly involved, for providing all the data we requested in a timely manner, and for sharing their perspectives with us. They earned our respect by treating different perspectives in a professional manner.

We are also indebted to the British and French Ministries of Defense for allowing us to visit their headquarters and to discuss, with the senior government leadership and their staffs, their experiences with production gaps, low-rate production, and production issues in general. We owe a debt to the management and staff of the DCN Shipyard at Brest, France; Vickers Shipbuilding and Engineering, Ltd., Barrow-in-Furness, England; Kvaerner Masa-Yards, Helsinki, Finland; Kvaerner Govan Shipyard, Glasgow, Scotland; Fincantieri Shipyard, Trieste, Italy; and Chantiers de l'Atlantique, Paris, France, for sharing with us their innovative organizational, management, engineering, and technical approaches to improving the quality of large, complex ships while driving down costs.

This broad-based participation made possible the analysis described here.

Finally, we wish to thank our RAND colleagues Jeff Isaacson and C. Richard Neu. Their thoughtful reviews occasioned many changes that improved the clarity of the report.

ABBREVIATIONS AND ACRONYMS

ABS	American Bureau of Shipping
AC	Allis Chalmers
ACCMP	*Aircraft Carrier Continuous Maintenance Program* (Manual)
AEW	Airborne early warning
AOR	Area of responsibility
AP	Advance procurement
ASS	Auxiliary support ship
ASW	Anti-submarine warfare
B&W	Babcock and Wilcox
BA	British Aerospace
BFM	Barry, Frank, & Murray
CAD/CAM	Computer-assisted design/computer-assisted manufacture
CAIV	Cost as an independent variable
CE	Combustion Engineering
CENTCOM	Central Command
CFE	Contractor-furnished equipment
CINC	Commander in chief
COTS	Commercial off-the-shelf (products)
CTOL	Conventional take-off and landing
CV	Carrier vessel (since mid-1970s)
CVA	Carrier vessel, attack
CVN	Carrier vessel, nuclear
CVS	Carrier vessel, anti-submarine
CW	Curtiss-Wright
DCN	Directorate for Naval Construction (in French)
DDG	Guided-missile destroyer
DGA	Delegation for Armament
DPIA	Docking planned incremental availability
EDSRA	Extended docking SRA
ESRA	Extended SRA
FASP	Flexible automated steel prefabrication
FICA	Federal Insurance Contributions Act

FW	Foster Wheeler
GE	General Electric
GFE	Government-furnished equipment
GOCO	Government-owned, commercially operated
HM&E	Hull, mechanical, and electrical
HVAC	Heating, ventilation, and air conditioning
IMO	International Maritime Organization
IMP	Incremental Maintenance Program
ITT	Invitation to Tender (British equivalent of Request for Proposal)
JUMPS	Joint Uniform Military Pay System
KGL	Kvaerner Govan Limited
LMTDS	Lockheed Martin Tactical Defense Systems
LNG	Liquid natural gas (carrier)
LPD	Landing platform–dock
LPG	Liquid petroleum gas
LPH	Landing platform–helicopter
lt	Long tons
M&C	Metals and Controls
M-S	Marvel-Schelber
MARC	Arctic Technology Center
MoD	Ministry of Defense (United Kingdom)
MOI	Ministry of Industry
MPN	Military personnel, Navy (budget category)
MRCs	Multiregional conflicts
MTW	Major theater war
NAO	National Audit Office (United Kingdom)
NASA	National Aeronautics and Space Administration
NAVSEA	Naval Sea Systems Command
NDRI	National Defense Research Institute
NED	Nuclear Equipment Division (BWX Technology)
NNS	Newport News Shipbuilding
NPV	Net present value
NSSN	New attack submarine
O&MN	Operations and maintenance, Navy (budget category)
O&S	Operating and support
OPNAV	Office of the Chief of Naval Operations
OPTEMPO	Operational tempo
OSD	Office of the Secretary of Defense
PCC	Precision Components Corporation
PERA-CV	Planning, Engineering, Repairs, Alterations—Aircraft Carriers
PERSTEMPO	Personnel tempo

PIA	Planned incremental availabilities
PSA	Postshakedown availability
PWR	Pressurized water reactors (nuclear)
R/LSI	Royal/Lear Siegler Inc.
R&D	Research and development
RCOH	Refueling/complex overhaul
RFP	Request for Proposal
SCN	Ship construction, Navy (budget cagetory)
SGLI	Servicemen's Group Life Insurance
SLEP	Service life extension program
s.h.p.	Shaft horsepower
SRA	Selected restricted availability
SSBN	Ballistic-missile submarine, nuclear-powered
std	Standard
STOVL	Short take-off/vertical landing
SUPSHIP	Supervisor of Shipbuilding
SW	Struthers Wells
UNC	United Nuclear Corporation
U.S.	United States
USA	United States Army
USN	United States Navy
USWA	United Steelworkers of America
VAMOSC	Visibility and Management of Operating and Support Costs (database)
VSEL	Vickers Engineering and Shipbuilding Limited
VSTOL	Vertical/short take-off and landing

Chapter One

INTRODUCTION

Proposing to build a nuclear aircraft carrier attracts a lot of attention. These warships are expensive, costing about $5 billion to construct. The largest warships built, they displace about 100,000 tons,[1] have a flight-deck area of almost 5 acres, and are nearly as long as the Empire State Building is tall. As high as a 24-story building from keel to mast, they accommodate over 5,000 Navy personnel for months at a time. They are expected to operate safely for decades—and, of course, to survive and function as fully as possible in crisis and conflict.

Dwarfing the magnitude of these gargantuan specifications are the industrial capability and capacity required to construct just one carrier. The time required to build a carrier ranges from 6 to 10 years and involves millions of direct labor hours just in the shipyard. A maximum of about 7,000 shipyard laborers work on the ship at one time. Over 1,000 vendors provide a wide range of components, as well.

The current carrier force structure includes 12 aircraft carriers. Four of these carriers are conventionally powered, and eight are nuclear-powered. Since 1968, the United States has built only nuclear-powered carriers[2]—and those at fairly regular intervals. In 1996, at the outset of this study, the Navy's plan called for construction to start on the last Nimitz-class ship, CVN 77, in the year 2002, with completion in 2008.[3]

---

[1]It is well to define the terms of tonnage measurement of vessels, since the terms vary for the type of vessel. For warships, the term is *displacement tonnage*—the volume of water displaced by the hull beneath the waterline (divided by 35 cu ft/ton). For passenger ships, the term is *gross tonnage*—a measure of the total volume of enclosed spaces on the ship (divided by 100 cu ft/ton). And for tankers and bulk cargo ships, the term is *deadweight tonnage*—a measure of the total volume of the ship dedicated to carrying cargo (divided by 35 cu ft/ton). These definitions are important, because some 100,000-gross-ton passenger ships have only about one-third the displacement of a 100,000-ton carrier.

[2]Delivered in 1968, *John F. Kennedy*, CV 67, was the last nonnuclear carrier.

[3]Construction start is preceded by a year or two of engineering work.

That schedule results in a 7-year gap since construction started on CVN 76, which was authorized in fiscal year 1995 (FY95). This gap exceeds any construction interval between individual carriers in the past 30 years. The longest previous gap was six years, between USS *Carl Vinson* (CVN 70, 1974) and USS *Theodore Roosevelt* (CVN 71, 1980). *Theodore Roosevelt* required more man-hours to complete than did earlier carriers, in part because of turnover and skill deterioration of the workforce.[4] It produced higher vendor costs than any other carrier of the Nimitz class built to date.

The Navy was concerned that a longer gap might cause such costs to be accrued again. At the same time, it had to consider the following:

- The debate about postponing the construction start of CVN 77 for a few years
- The possibility that budget constraints might cause construction of CVN 77 to be bypassed completely to make way for the first ship of the next class, designated CVX, which is now scheduled for a construction start in 2006
- The fact that the average age of U.S. conventional carriers in 2008 will be 44 years. As these ships age, they become increasingly expensive and difficult to maintain.

Motivated by these considerations, the Navy asked RAND to perform an independent, quantitative analysis of the aircraft carrier industrial base, focusing on CVN 77. We sought to determine whether the industrial base was adequate to support future carrier production. Additionally, we sought answers to the following questions:

1. What constraints does the need to meet carrier force requirements place on start dates for CVN 77? To answer this question, we graphically demonstrate the relations among fleet size, ship retirement age, and the schedule of new-ship construction (in Chapter Three).
2. Of those start dates responsive to force requirements, which permits the most economical carrier construction? Do longer gaps entail risks? Here, we report the results of a mathematical model that determines, for any start date, the least-cost shipyard workforce profile, given the other work planned for the yard building the carrier (in Chapter Four).
3. What are the implications of different start dates for the survival of vendors supplying carrier components to the shipyard? What are the implications for

[4]Moving carrier production to another part of the shipyard, a change in production sequence, and an accelerated production schedule also contributed to the high man-hours.

the cost of those components? We answer these questions separately for the suppliers of nuclear and nonnuclear components (in Chapters Five and Six, respectively).

4. Are there any production processes or technologies not now employed in Navy shipbuilding that could permit significant savings in carrier life-cycle costs? If so, some research and development will be required to adapt such processes and technologies to aircraft-carrier construction and operation. How much R&D do the potential savings justify? We found the answer to the first question through interviews with manufacturers of large, complex ships for the commercial sector. We derived the second answer by estimating key maintenance and personnel costs, then conservatively hypothesizing a certain percentage savings (in Chapter Seven).

Of course, no projection of future costs and other consequences can be made with certainty. The implications of varying CVN 77's start date are mediated by the timing of construction, maintenance, and retirement of other warships. For that timing, we have sought consistency with current Navy plans, because they appear at least as likely as any other future and because doing so should be of greatest utility to acquisition planners in the Navy and elsewhere. Assumptions specific to the various elements of our analysis are taken up in the relevant chapters. We stress here that if the flow of work through the shipyard and vendors varies significantly from that now planned, the results of this analysis could also vary, not only in the size of the predicted effects but also in their direction.

Chapter Two

AIRCRAFT CARRIERS AND THE CARRIER INDUSTRIAL BASE

In this report, we take as a baseline a fleet of 12 aircraft carriers whose primary role remains as it has been,[1] and we assume that the basic design features of the current class of carriers will apply to CVN 77, the last ship of that class. The past and potential future evolution of carrier force structure, roles, and design thus do not much affect our analysis of industrial-base issues. Nevertheless, some knowledge of that evolution provides an important context for our industrial-base analysis and the implications to be drawn from it. In this chapter, we offer background on the role, force structure, and design of carriers and on the carrier industrial base.

THE CARRIER'S ROLE

As with the rest of the U.S. military, aircraft carriers exist to support the National Military Strategy of the United States—a strategy that has, of course, evolved over the past 50 years, as have perceptions of threats to U.S. national security. The result has been a sorting out of roles among elements of the force structure.

The Early Days[2]

In the early part of this century, the Navy's primary offensive weapon was the battleship's heavy guns. The first carriers—those built before World War II—operated as an adjunct to the battle fleet, providing the battleships with such vital services as reconnaissance and spotting, and controlling the air over the

[1]The Navy has been aggressive in thinking about new paradigms for aircraft carriers. For a thorough discussion, see Jacquelyn K. Davis, *CVX: A Smart Carrier for the New Era*, Washington, D.C.: Brassey's, 1998.

[2]The United States has produced (i.e., built or converted and launched) a total of 64 carriers in 15 different classes. See Appendix A.

gunnery engagement. The strike power of a carrier was, at first, very much a secondary asset.[3]

This situation began to change as carrier forces were modernized. The new carrier emphasis was demonstrated by the British at Taranto and the Japanese at Pearl Harbor. The first carrier battles in the Coral Sea and northwest of Midway Island finalized this reorientation in thinking. In the Midway battle in particular, the Japanese carrier force was smashed without a heavy gun being fired.[4] Within three weeks, the Japanese canceled battleship construction and implemented a new construction and conversion program that emphasized carriers.[5] The United States already had 23 carriers under construction; within one month after Midway, Congress authorized 13 more, and 10 of those were ordered from shipyards within 30 days.

Postwar Adaptation

Following World War II, the Navy was forced to adapt to a changing world environment. Whereas wartime experience had centered on defeating the Japanese fleet in blue-water engagements, the Soviet Union was a classic Continental power that presented no significant naval threat. This difference left the Navy carrier leadership in search of a mission. To this end, the service tried to assimilate the new nuclear-strike role. The outcome of the resulting debate has shaped the aircraft carrier's function to this day.

The Navy's solution to the strategic nuclear-strike problem—a carrier that could conduct a heavy, primarily nuclear, bombing attack against land targets—was USS *United States* (CVA 58), construction of which was started in 1949.[6] Envisioned as the platform for the heaviest long-range jet aircraft that could practicably operate from a carrier deck, *United States* was to be dramatically larger than previous carriers and was designed for this new strike role. Because of that role, she would not replace existing carriers but would form the centerpiece of strike groups including conventional carriers operating multirole air wings—a specialization representing a departure from the aircraft

---

[3]For a detailed discussion, see Norman Friedman, *U.S. Aircraft Carriers: An Illustrated Design History*, Annapolis, Md.: Naval Institute Press, 1983.

[4]Karl Lautenschlager, *Technology and the Evolution of Naval Warfare 1851–2001*, Washington D.C.: National Academy Press, 1984.

[5]The Yamato class was planned to consist of seven battleships. *Yamato* was commissioned just after the start of the Pacific War in 1941, *Musashi* in the middle of 1942, and work on *Shinano* and *No. 111* ceased at the outbreak of World War II. After the Battle of Midway, *Shinano* was redesigned as an aircraft carrier. *No. 111* and the remaining three ships were canceled.

[6]Roger Chesneau, *Aircraft Carriers of the World, 1914 to the Present: An Illustrated Encyclopedia*, Annapolis, Md.: Naval Institute Press, 1984.

carrier's major strengths of flexibility and effectiveness over a wide range of warfighting scenarios.

At the time, the Navy was willing to sacrifice this flexibility to gain a major share of the dominant nuclear-strike mission. This role was controversial because it directly opposed Air Force development of a large force of intercontinental bombers, which, for many political, technical, and budgetary reasons, won out in the end.[7] *United States* was canceled by Secretary of Defense Louis Johnson in April 1949, eight days after the keel was laid. Ironically, the Navy did not lose a nuclear-attack capability; it gained one by developing such heavy-attack aircraft as the AJ-1 Savage, A-3 Skywarrior, and A-5 Vigilante. Thus, although the role of operating nuclear-strike aircraft was assigned primarily to the Air Force, the Navy gained a limited carrier-based strategic-strike capability.

The cancellation of *United States* led the Navy to de-emphasize the specialized, heavy-attack carrier and focus on acquiring and employing more-flexible and more-adaptable carriers, aircraft, and doctrine: "Between 1946 and 1950, the concept of future carrier operations shifted from strategic strikes by a small number of carrier-based heavy bombers to tactical air strikes by a much larger group of smaller aircraft."[8] The outbreak of the Korean War soon confirmed the value of this tactical orientation.

Because of geography and the rapid advance of the communist forces, allied airpower early in the conflict was primarily carrier-based: Carriers provided rapid response and the ability to generate a high number of sorties while remaining relatively immune to land-based threats. The Korean experience thus provided the model of the multirole, tactically oriented, forward-deployed carrier strike force, a force that was to be called upon repeatedly in the coming decades.

Carriers provided a major portion of U.S. airpower during the Vietnam conflict; they have performed combat operations many times since, most notably in Grenada and Lebanon (1983) and Libya (1986), and in Operation Desert Storm (1991). Despite the lack of a credible, ocean-going naval threat to American interests, carrier-based tactical airpower came of age during the Cold War as the major, and sometimes sole, instrument of American power projection.

---

[7]Of course, the Navy did acquire a strategic nuclear-strike role through submarine-launched ballistic missiles.

[8]Friedman, 1983, p. 255.

Forward Presence and Crisis Response[9]

The current National Military Strategy stresses overseas presence and power projection, to enable promotion of stability and, when necessary, defeat of adversaries:

> Forward deployed naval expeditionary forces can respond immediately to a crisis . . . and through prompt action help halt an enemy offensive and enable the flow of follow-on . . . contingents. By ensuring freedom of the seas and controlling strategic choke points, naval forces provide strategic freedom of maneuver and enhance deployment and sustainment of joint forces in theater. Air forces maintain control of the skies, helping to destroy the enemy's ability to wage war, providing sustained, precise firepower, and numerous tactical and operational advantages while facilitating land and naval maneuver.[10]

The aircraft carrier combines peacetime engagement, deterrence, and war-fighting capabilities in one integrated package. It is often the initial power-projection and enabling asset when hostilities threaten or occur. Short of hostilities, the presence of an aircraft carrier makes a powerful political statement and presents a credible military threat to enemies, along with offering support for allies.[11] As was made clear in the 1991 Gulf War, crisis deployments of land-based air and ground forces to allied countries are necessary; however, positioning them takes time and obviously requires the host nation's cooperation.[12] To a government whose populations or neighbors may be opposed to the operations of U.S. land-based aircraft from its territory, carrier battle groups stationed offshore in international waters can satisfactorily balance the demands of internal politics and external defense.

DESIGN EVOLUTION

Despite its cancellation, the *United States* program shaped the design of future carriers. Relative to the carriers built in the first half of the century, *Forrestal* and ships that followed were distinguished by

- increased size, including the ability to operate increasingly fast, heavy, and capable aircraft throughout the life of the ship

---

[9]Data on U.S. military crisis response are presented in Appendix B.

[10]John Shalikashvili, *National Military Strategy of the United States*, Washington, D.C.: Joint Chiefs of Staff, 1995, p. 14.

[11]*Independence*, which is conventionally powered, is homeported in Tokyo Harbor, the only U.S. carrier to be based permanently outside the United States.

[12]The United States has built 172 overseas bases since World War II. It has access to 24 of those today. (Davis, 1998, p. 8.)

- enlarged capacities to store fuel, ordnance, and other supplies, increasing the carrier's ability to sustain long periods of combat operations
- improved survivability
- improved all-weather operations and seakeeping ability
- increased longevity.

Forrestal—The First Modern Aircraft Carrier Class

The Korean War of 1950–1953 was a sharp reminder that conventional military conflict could erupt and challenge U.S. interests. The war saw extensive use of carriers and prompted construction of a new class of larger vessels: the Forrestal class.[13] The increased size of this class was a result of prudently modifying *United States'* design to take into account rapidly evolving jet-aircraft development. Four ships of this class were authorized in successive years beginning in FY52.

Before the Forrestal class was constructed, the evolving performance of aircraft had been outstripping the carrier's ability to handle them safely. Techniques of shipboard launch and recovery represented a major obstacle to the introduction of modern jet aircraft. It was not until 1955, when the United Kingdom's HMS *Ark Royal* went into service, that all three features essential for jet-age aviation appeared on a carrier:[14]

- Steam catapults that could accelerate heavier, swept-wing jets to the higher airspeeds they required to become airborne
- An angled flight deck, which permitted safe recovery of jet aircraft
- An optical landing system to show the pilot the proper glide slope for a safe landing.[15]

Later that same year, *Forrestal* was commissioned with these same characteristics. *Forrestal* could store 1,800 tons of ordnance and 750,000 gallons of aircraft fuel. It was also the first carrier in which the flight deck was an integral structural member of the hull. This design allowed the deck to support 75,000-lb A-3 heavy-attack jets. Twenty years later, this structure was sound enough to

[13]During the 1950s, World War II–era Essex-class ships were also brought out of retirement and modified by installing new flight decks to handle jet aircraft. However, these carriers were increasingly relegated to anti-submarine warfare operations that did not require handling jet fighter aircraft.

[14]For a list of carriers currently in operation by all the world's navies, see Appendix C.

[15]Lautenschlager, 1984, p. 49.

handle the even-heavier F-14. By contrast, the ships of the 1940s-era Midway class were incapable of operating F-14s and, because of their smaller size, were often restricted by inclement weather.

The enhanced survivability of larger vessels is epitomized by *Forrestal* herself. Although seriously damaged by the famous fire and explosions that occurred on board in July 1967 off Vietnam, "the carrier survived. Whether a ship of smaller dimensions would have done so is open to speculation."[16] Finally, their size and the forethought with which these ships were constructed enabled them to spend their service lives without major modifications. Neither the Forrestals nor subsequent U.S. carriers have undergone structural alteration as significant as that affecting the Midway and Essex classes.[17]

Nuclear Propulsion

Introduction of nuclear power was the next defining step in carrier evolution. Envisioned originally in the late 1940s, nuclear-power-plant technology was not sufficiently advanced to be incorporated in the Forrestals. Finally authorized in 1956, the first nuclear-powered carrier, USS *Enterprise* (CVN 65), was significantly larger, more complex, and costlier than previous vessels. Her displacement was a maximum 94,000 tons—compared with the 81,000 tons characteristic of the "improved" Forrestal-class ships (CVs 63, 64, 66)—and allowed room for approximately 90 percent more aviation fuel (up to 2.75 million gallons) and 50 percent more ordnance (up to 2,500 tons).

The main advantage of nuclear power was virtually unlimited endurance, as displayed through *Enterprise*'s highly publicized round-the-world cruise. The power plant thus permitted *Enterprise* to sustain operations even longer than had previous large carriers. It demonstrated such sustainment initially off Vietnam and again during 1979–1981 deployments to the Indian Ocean in response to crises in Iran and Afghanistan. The latter operations are notable because they were accomplished with little logistics support in a region rarely visited by U.S. forces until then.

The Nimitz class included improvements over the one-of-a-kind *Enterprise.* By substituting two more-efficient nuclear reactors for *Enterprise's* eight, the Nimitz class achieved even greater storage for aviation fuel, ordnance, and spare parts.

---

[16]Chesneau, 1984, p. 268.

[17]Chesneau, 1984, p. 267.

Nuclear-powered carriers offer advantages besides storage and endurance, but those advantages may not be so obvious at first. Two of the most significant are speed and acceleration. Nuclear reactors can quickly impart high speed to the carrier's massive hull, permitting great operational flexibility; the ability to sustain that speed heightens the ship's responsiveness. Increased speed and endurance also make the nuclear-powered carrier more difficult to locate and target, and thus make it less vulnerable to modern submarines and cruise missiles. Another set of advantages accrues from the electrical and steam-power reserves of a nuclear plant. The substantial amounts of steam power required to launch increasingly heavy aircraft with safety can be had much more reliably with nuclear power than with fossil fuels. And the electrical demand of a carrier, already huge, will only increase with the increasing importance of electronics, computers, and radars.[18]

Although at the outset technically difficult and expensive to achieve,[19] the use of nuclear propulsion on aircraft carriers has proven its worth militarily and has generated significant economies over the life of the ships.

The Next Class and the Disadvantages of the Nimitz

The next carrier to be built after CVN 77 is planned to be the first ship of a new class, designated CVX.[20] The design of the CVX has yet to be determined, even in broad outline. It may well include only evolutionary improvements to the tested, long-prevailing design concept of large nuclear carriers. However, Nimitz-class carriers are not without their disadvantages, and the Navy may consider smaller, nonnuclear designs—although, historically, warship types have not been scaled back in size.

We do not wish to speculate here about whether historical precedent will apply. Instead, we take this opportunity to point out some of the issues concerning large, Nimitz-class carriers. Cost is, of course, chief among the disadvantages. Whereas the calculus of cost-effectiveness is complex and may favor the Nimitz class, there is no question that periodically replacing large Nimitz-class carriers with newer, similar ships takes a big bite out of the Navy's ship-construction resources. (Typically, a ship is funded in large part, and often completely, from a

---

[18]For a full discussion of the various advantages of nuclear power for carrier design and operations, see Friedman, 1983, p. 309.

[19]Building nuclear-powered vessels for the Navy is far different from building ships to standards for conventional power.

[20]For an interesting discussion of this design in relation to aviation at sea, see Reuven Leopold, *Sea-Based Aviation and the Next U.S. Aircraft Carrier Design: The CVX*, Cambridge, Mass.: Massachusetts Institute of Technology, Center for International Studies, MIT Security Studies Program, January 1998.

single year's construction budget (except for advance procurement [AP] of nuclear components.)

Of course, Nimitz-class carriers also provide greater military capability, so a trade-off must be reached between cost and military value. Also, as will become apparent in Chapter Three, construction of large, Nimitz-class ships with ship funding confined to one year limits the Navy's flexibility in spreading work over shipyards and over time.

Other disadvantages of Nimitz-class carriers may be more perceived than real; they fall into three categories: (1) the disposal of radioactive waste (including spent nuclear fuel) produced during ship operation, (2) the acceptance of nuclear-powered warships in foreign ports, and (3) the potential environmental impacts if a naval reactor accident occurs. We deal briefly with each in turn here.

The Navy has a long history of safely disposing of the nuclear waste it produces from nuclear-powered warship operation and servicing. Indeed, the amount of low-level radioactive waste it generates constitutes less than 10 percent of what is produced by commercial nuclear-power plants in the United States. Existing and projected disposal capacity, required for commercial nuclear-power-plant operation, is more than sufficient for the Navy's needs.

With respect to spent nuclear fuel, a Nimitz-class carrier generates several metric tons over its lifetime—a very small amount compared with what commercial nuclear reactors produce: a projected 85,000 metric tons (heavy metal) by the year 2035. By that same point, the Navy expects to have generated about 65 metric tons of spent fuel from all of its nuclear-powered warships. All spent fuel is planned for disposal in a geologic repository. While the final costs of disposal have not yet been established, the Navy expects the disposal costs for the spent fuel created by a Nimitz-class carrier over its lifetime to be less than $20 million—a very small fraction of the life-cycle cost of the ship.

The Navy's nuclear-powered warships have a long history of gaining access to foreign ports, owing to their long-standing record of safety and environmental protection. Nuclear-powered warships visit over 150 ports in 50 foreign countries and dependencies, including major industrialized countries such as Japan, Germany, Great Britain, France, and Canada. While it cannot be denied that a reactor accident may adversely affect the Navy's ability to enter some foreign ports, the Navy is keenly aware of that potential. Its record demonstrates that avoiding such a problem is the focal point of its "safety-first" mentality.

The likelihood of a reactor accident—the release of fission products from the reactor—is extremely small, because the reactors in Nimitz-class carriers are designed to military standards for shock, battle damage, and reliability. Navy nuclear-powered warships have accumulated over 4,800 reactor-years of operation, and have steamed over 110 million miles, without such an event occurring.

Even if an accident occurs, the impacts on the environment and on the public are expected to be small, for four reasons:

- The power of a Nimitz-class reactor is rated at less than 20 percent that of a typical commercial nuclear power plant.
- Unlike commercial plants, which typically operate continuously at their maximum-rated power to generate revenues from electrical-energy production, naval reactors usually operate much below their maximum power ratings, since there is no need to proceed continuously at maximum speed. (The amount of radioactive material available to be dispersed in the event of an accident is much lower if the reactor is operating at low power before the accident.)
- When a carrier is in port, its reactors are usually shut down or are operating at very low power; the ship uses shore power for its "hotel" functions—housing, laundering, food preparation, etc.
- Unlike a land-based power plant, a ship is mobile: It can be moved away from populated areas if a problem occurs.

FORCE STRUCTURE

The structure of a carrier force is now largely determined by the ships' role in supporting America's national security objectives and takes into account the design of the ships currently in the force. (Figure 2.1 traces U.S. carrier force structure from 1950 to 1997, by hull number and class.[21]) When the United States began building a carrier force, this was not the case. Force size was determined, instead, by the 1922 Washington Naval Treaty. Freed from treaty limitations by World War II, the carrier force quickly expanded. Most carriers built during World War II were of the 35,000-ton Essex class. Extensively modified, many of these served through most of the Cold War as both attack (CVA) and anti-submarine (CVS) carriers alongside the larger Midways, Forrestals, and *Enterprise*. However, as early as 1947, a Cold War policy-

---

[21]Full service lives are shown for all ships operating since 1950, so the figure extends back to 1943 to accommodate these. However, the figure is not an accurate representation of force structure in the 1940s, because some ships in that force structure were no longer operating in 1950 and are thus not included. If a ship was commissioned by June 30 or decommissioned after June 30, it is counted as being in operation for that year.

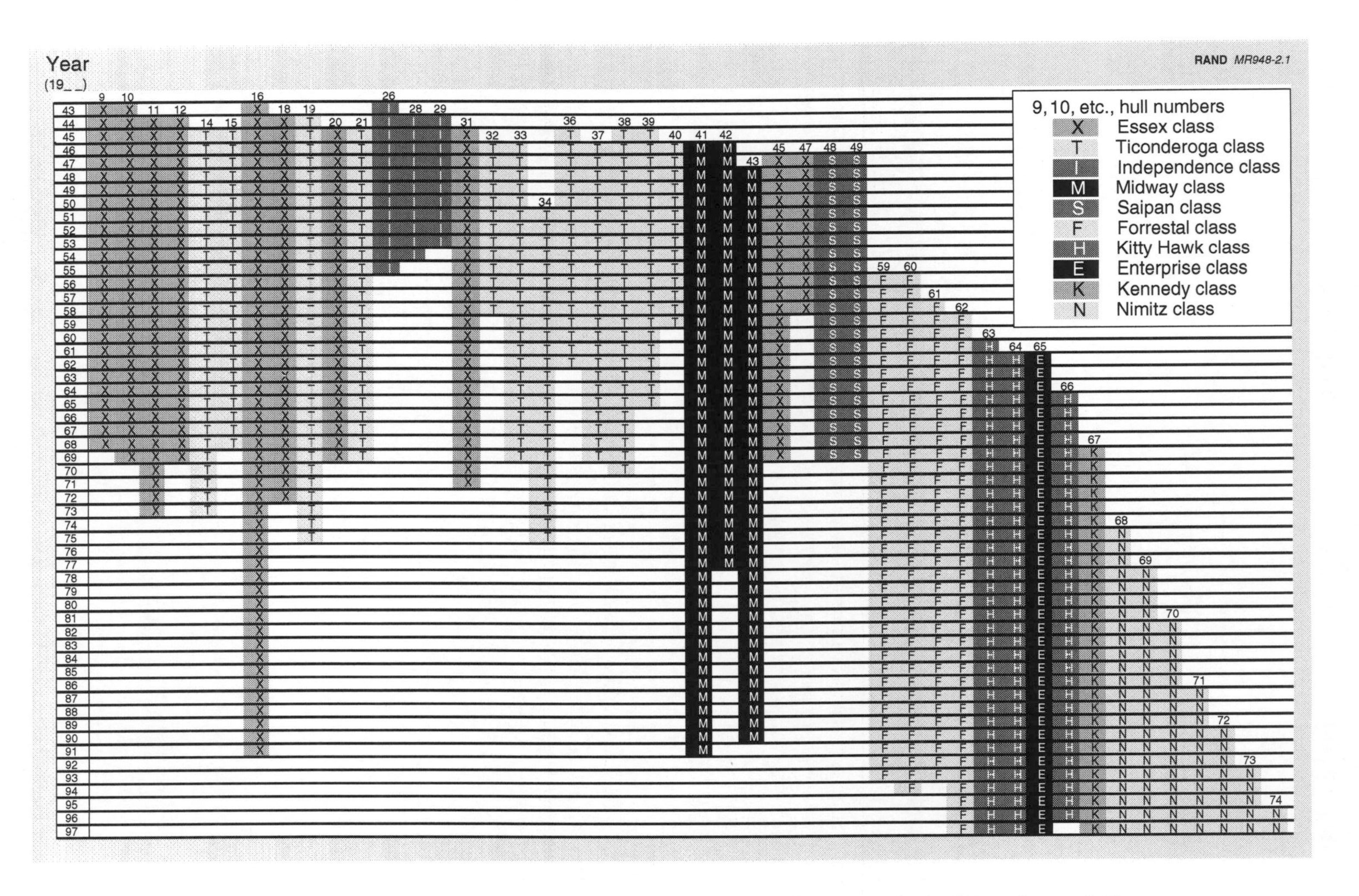

Figure 2.1—Service Lives of U.S. Aircraft Carriers in Operation Since 1950, by Hull Number and Class

planning document called for 12 CVAs, and the number of fleet carriers—those deploying fighters and strike aircraft—has generally held between 12 and 15 (Figure 2.2). Thus, although there was an apparent surfeit of carriers in the 1950s and 1960s, many were less capable than the larger, newer ships.[22]

With the prospect of large-scale deactivations and new-carrier construction in the late 1960s, carrier force structure became an issue. In light of the Vietnam War experience, Secretary of Defense Robert McNamara in 1966 increased the 1970s force structure from a proposed 13 attack carriers to 15.[23] This number remained relatively constant until post–Cold War restructuring brought the number down to 12. All along, as more-modern ships replaced aging ones, the carrier force changed in composition as well as in number (see Figure 2.1).

Why a 12-ship fleet? Currently, carrier force structure is based primarily on support of the commanders in chief of U.S. forces in the Western Pacific Ocean, Europe/Mediterranean Sea, and Indian Ocean/Persian Gulf. Maintaining a continuous carrier presence in each of these three areas would require approximately 15 ships—the rationale for the Cold War policy. This 5-to-1 ratio allows for maintenance, training and predeployment exercises, and personnel time in home port,[24] and it accounts for transit time to operational areas from the West Coast of the United States. In the post–Cold War era, limited gaps in carrier presence are deemed acceptable, so current national security objectives are regarded as satisfied with 12 aircraft carriers. (Table 2.1 delineates the coverage gaps experienced with fleets smaller than 15 ships.)

Note that, in discussing the adequacy of a 12-ship fleet, we have been assuming normal peacetime operations only. More-serious contingencies such as humanitarian intervention and peace enforcement may require more than one carrier on-scene. Examples from 1996 alone are *Nimitz* and *Independence* off Taiwan, and *Enterprise* and *Carl Vinson* in the Persian Gulf. Libyan operations in March 1986 used three Atlantic Fleet carriers in the Mediterranean at one time. A major theater war (MTW) is estimated to require four or five carriers—six were ultimately involved in Operation Desert Storm—and the National Military Strategy envisions response to two MTWs simultaneously. Thus, in meeting such contingencies, something must give: presence in other areas, maintenance schedules, or sailor time at home port.

---

[22]The majority of the Essex-class ships were decommissioned in the late 1960s and early 1970s.

[23]Friedman, 1983, p. 318. Previously separate anti-submarine and attack functions were combined in carrier air wings by the mid-1970s. All carriers were redesignated "CV."

[24]Chief of Naval Operations personnel policy established in 1986 that deployments shall not exceed six months in length, and that personnel shall remain home for 12 months after completing a 6-month deployment.

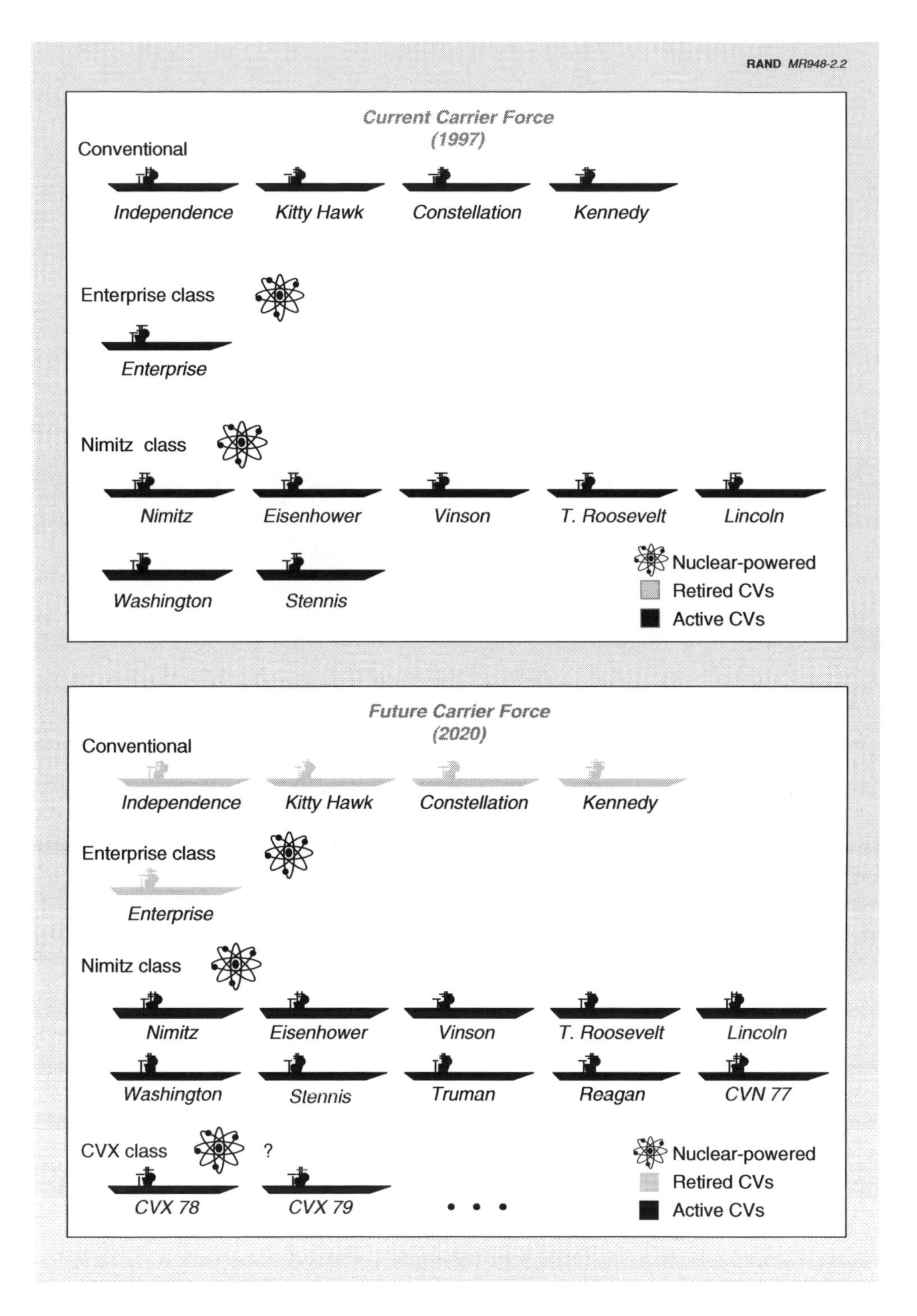

Figure 2.2—Current and Future Carrier Force Structures

Table 2.1

Continuous Carrier Forward-Presence Coverage in Three Key Regions

	Region		
Fleet Size	Mediterranean	Western Pacific	North Arabian Sea
	Indian Ocean/Persian Gulf Focus		
15	100%	100%	100%
12	81% (2.3-month gap)	80% (2.4-month gap)	100%
10	67% (4.0-month gap)	65% (4.2-month gap)	100%
	Mediterranean Focus		
15	100%	100%	100%
12	100%	87% (1.6-month gap)	79% (2.5-month gap)
10	100%	70% (3.6-month gap)	73% (3.2-month gap)

SOURCE: Director, Assessment Division (OPNAV Staff N81), April 29, 1993.

NOTES: *Gap* means there is no carrier in the unified commander in chief's (CINC's) area of responsibility (AOR). Gaps are months per year. The plans shown meet personnel-tempo and operational-tempo objectives; they account for both routine and longer-term (nuclear refueling) overhauls and other maintenance requirements.

The need for crisis response thus represents a continuing demand on the 12 aircraft carriers remaining in the U.S. fleet. Figure 2.3 suggests that, if there was no excess of carriers before the end of the Cold War, there may not be an excess now.

Of the various factors underlying the 5-to-1 fleet-size-to-on-station ratio, ship maintenance is particularly important and plays a big role in our analysis. Figure 2.4 illustrates the nominal breakdown of a Nimitz-class carrier's life-cycle activities as outlined in the CVN 68 *Incremental Maintenance Program* (IMP).[25] Here "Maintenance" refers only to scheduled shipyard activities (it includes nuclear refueling), "Deployment" time is defined as long missions overseas, and "Training" is all other underway, homeport, and upkeep periods. This allocation of life-cycle activities is designed to achieve the objectives for personnel tempo (PERSTEMPO) and operational tempo (OPTEMPO);[26] improve the capability to meet large, sudden deployment demands; and reduce the probability of having more than one ship in maintenance at a time. In other words, it is designed to keep as many carriers operational as possible while

[25]Planning, Engineering, Repairs, Alterations—Aircraft Carriers (PERA-CV), *Incremental Maintenance Program for CVN-68 Class Aircraft Carriers*, Bremerton, Wash., January 1, 1997, p. 1-2.

[26]PERSTEMPO is an expression of the ratio between the amount of time during a ship's operating period that personnel must spend in their home ports and the amount of time under way or in foreign ports. The emphasis has been to ensure that PERSTEMPO meets established objectives (see footnote 24). OPTEMPO refers to the frequency and duration of at-sea operations and training of all naval (Navy and Marine) forces.

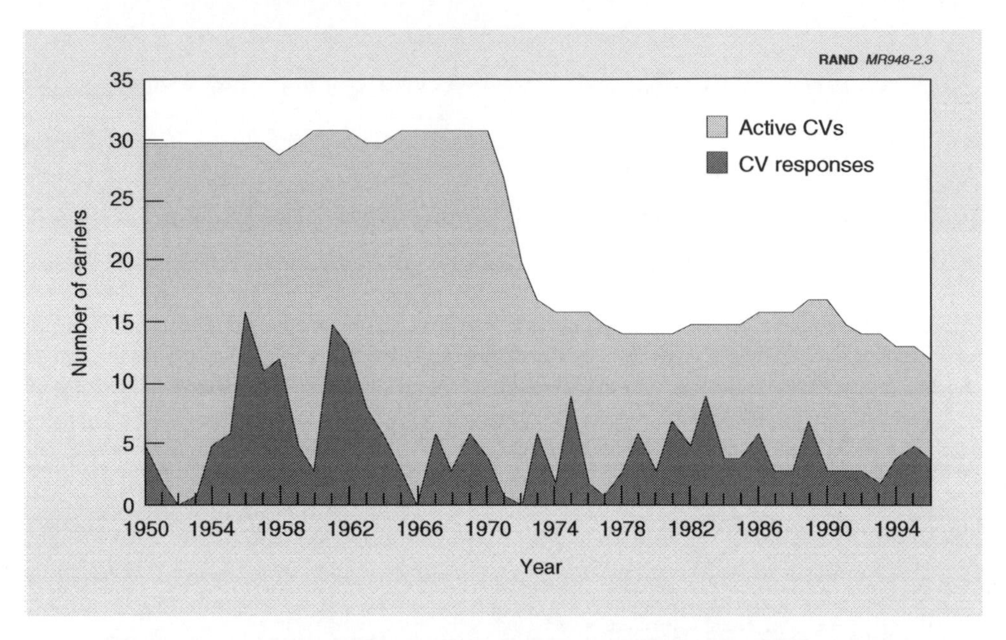

Figure 2.3—Carrier Response in Relation to Number of Carriers in Fleet

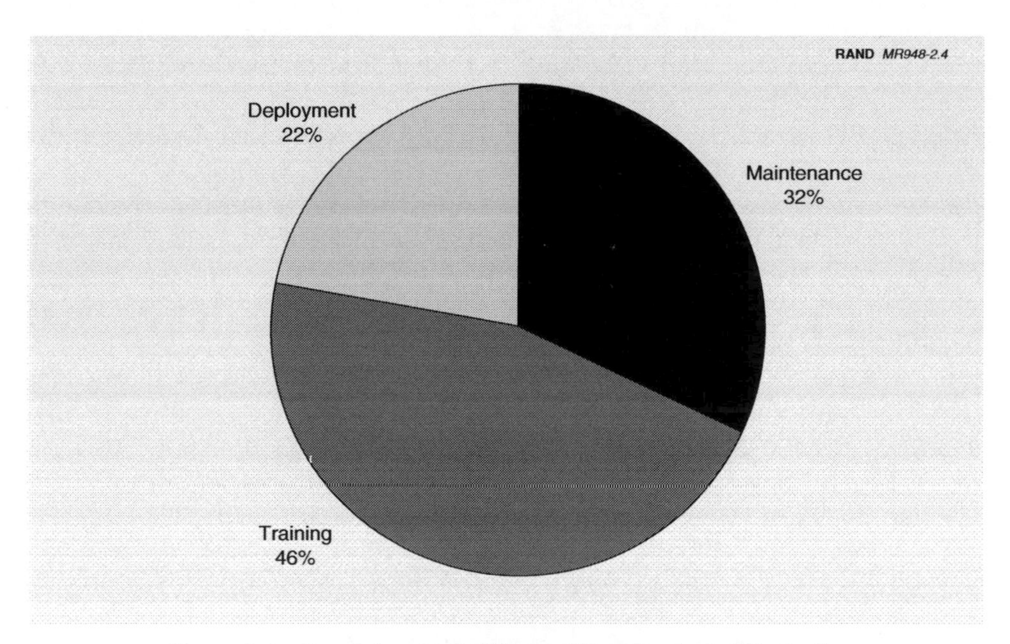

Figure 2.4—Breakdown of a Nimitz-Class Carrier's Life Cycle

maintaining the ships' material condition and maximizing their useful life. Just the same, under the plan, less than one-fourth of a carrier's 48-year life is actually spent deployed forward, on-station.

CHALLENGES TO AIRCRAFT CARRIERS

A number of critics have challenged the future value of aircraft carriers. These critics have made the following claims:

- There are other ways to project power. The United States could place greater emphasis on surface-ship- or submarine-launched cruise missiles or B-2 bombers instead of on aircraft carriers.
- Carriers are an expensive way to project power. An aircraft carrier requires several escort ships to provide defense against air, missile, and submarine attack. It requires auxiliary ships to deliver supplies and aviation fuel. Aircraft carriers today deploy approximately 50+ strike aircraft (F/A-18s and F-14s).
- Carriers represent too big a target. Loss of an aircraft carrier would be a major political blow to the United States and a tragic event in its own right. A carrier and the air wing deployed on the ship have a crew of approximately 5,000 persons.

The debate over such issues will continue, but the Navy and other military commanders have found the carrier to be a very flexible and valuable platform in responding to a crisis and to be key during a conflict.[27] Carriers deploy not just aircraft that can shoot down other aircraft or drop bombs and missiles on targets; they deploy surveillance aircraft that can provide control of a broad swath of sea, and they provide aircraft that can deploy mines or conduct anti-submarine operations. More recently, they have become *an integral part of peacekeeping operations.*

In the 1970s, the Navy extensively investigated smaller aircraft carriers—approximately one-half the size of the Nimitz-class vessels—under the assumption that smaller carriers would be cheaper than big ones. But the size of the ship limits the number of aircraft that can be deployed and their capabilities. A

---

[27]The Commander of Operations Desert Shield and Desert Storm, Army General H. Norman Schwarzkopf, pointed out that, in August 1990, the carriers *Eisenhower* and *Independence* were within range of Iraqi targets less than 48 hours after President George Bush issued a deployment order. In Schwarzkopf's words, "the Navy was the first military force to respond to the invasion, establishing immediate sea superiority. And the Navy was also the first air power on the scene. Both of these firsts deterred, indeed—I believe—stopped, Iraq from marching into Saudi Arabia." Speech to the 1991 graduating class of the United States Naval Academy, reprinted in *Proceedings*, U.S. Naval Institute, August 1991, p. 44.

smaller ship can deploy fewer, less-capable aircraft, and it could prove less survivable than a larger ship.

The issue of whether bombers and cruise missiles can be substituted for aircraft carriers is still being debated. Cruise missiles remain expensive—$1–$2 million each—and are fine for destroying high-value targets. But they can be shot down or deceived, and, once launched, cannot be recalled. Bombers can attack any point on the globe, but they need refueling support. Bombers can only project power; they cannot remain on-station to control airspace or effect sea control. Additionally, bombers reacting from the United States are not present to deter aggression or prevent crises from occurring. Both bombers and cruise missiles could play a role at the margin of the debate on future aircraft-carrier needs. But, barring a major technology or cost breakthrough, we do not envision either B-2s or cruise missiles fundamentally altering the outlook for carrier demand.

In some ways, carriers have become more important in the post–Cold War era. The United States sustains far fewer forces abroad today, and local basing issues limit the United States' ability to make on-the-ground deployments. The United States has focused its interests on many areas of the world that have less-sophisticated infrastructures than Europe and that may not support a large deployment of ground-based aircraft.

THE AIRCRAFT CARRIER INDUSTRIAL BASE

To help in understanding aircraft-carrier construction issues, we briefly characterize the current carrier industrial base.

America's commercial shipping and shipbuilding industry has declined dramatically since World War II. Whereas numerous public[28] and private shipyards once constructed naval vessels, including aircraft carriers, today only six commercial yards remain that can build major naval vessels. Of these, only one, Newport News Shipbuilding, constructs nuclear-powered and conventionally powered aircraft carriers. Electric Boat Corporation is also nuclear-capable but builds submarines exclusively.

Newport News Shipbuilding was founded in 1886 and is headquartered at Newport News, Virginia. This yard, a land-level facility located along the banks of the James River, occupies approximately 550 acres, and comprises seven

[28]*Public shipyards* are those yards operated by the Navy. At present, they are Norfolk, Portsmouth, Puget Sound, and Pearl Harbor Naval Shipyards. None constructs ships; they are used for repair and overhaul activities (all are nuclear-certified). Facilities at Mare Island and Long Beach, Calif.; Charleston, S.C.; and Philadelphia, Penn., have been, or are being, closed.

graving docks (dry docks), including the largest dry dock in the Western Hemisphere; a floating dry dock; two outfitting berths; and five outfitting piers. It has 17.5 acres of all-weather on-site steel-fabrication shops and maintains its own technical school to train apprentices for skilled-labor positions.

Newport News has generally built large surface warships, including carriers, battleships, cruisers, and destroyers, as well as submarines and commercial ships. Since 1960, its focus has been on nuclear-powered ships, principally aircraft carriers and submarines. It was the lead designer of the Los Angeles–class nuclear attack submarine program.

The yard's dominance in aircraft carrier construction is what concerns us here. Of the 64 fleet carriers built and launched for the U.S. Navy, 29, or 45 percent, have been constructed by Newport News. Most of those not built by Newport News were wartime Essex-class ships. More significant, 13 of 17 modern, large-deck carriers, including all nuclear-powered carriers, have been constructed by this one yard. Not only is Newport News the only yard currently geared for carrier construction, it is the only yard to build carriers for the past 36 years (see Figure 2.5). This seeming monopoly is due partially to the requirement for unique facilities that the construction of large carriers—and particularly complex, nuclear-powered ships—poses.

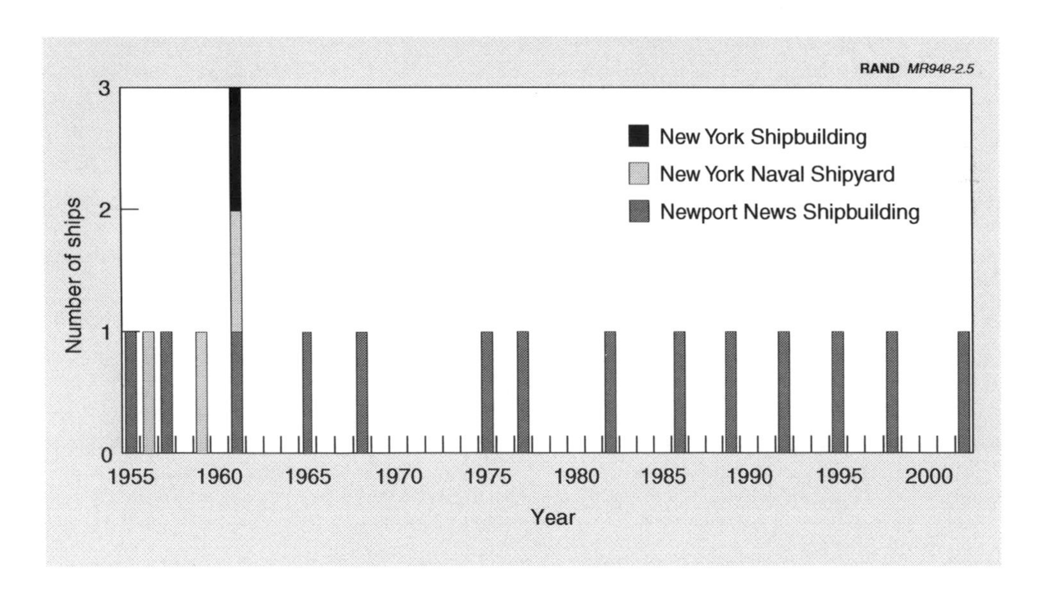

Figure 2.5—Carriers Completed Each Year, by Shipyard

Newport News has made and continues to make substantial capital investments to improve carrier shipbuilding processes and facilities, such as the automated steel factory and dry-dock extensions. Additional improvements are expected from lessons learned in NNS's ongoing commercial-ship-construction and overhaul projects. Two other commercial yards, Ingalls Shipbuilding, Inc., and Avondale, could be made capable of constructing *conventionally* powered aircraft carriers.[29]

---

[29]For more on commercial Navy shipbuilders, see Ronald O'Rourke, *Navy Major Shipbuilding Programs and Shipbuilders: Issues and Options for Congress*, Washington, D.C.: Congressional Research Service, September 24, 1996.

Chapter Three

HOW FORCE STRUCTURE OBJECTIVES CONSTRAIN CARRIER PRODUCTION SCHEDULES

In determining whether the shipbuilding industry can meet future Navy aircraft-carrier needs and how different carrier-construction schedules might lead to different overall acquisition costs, we must project what we take to be future Navy carrier needs: what must be built by when. It is only within these constraints that we can meaningfully estimate the cost implications of various production schedules. In this chapter, we explore those demands and constraints and, in so doing, identify the critical factors that link construction schedules with force composition. We also determine which new-carrier construction schedules would support force-structure objectives.

MAJOR FACTORS AFFECTING FORCE COMPOSITION

Two factors control the size of the carrier force as it evolves over time:

- The age of a ship at which replacement is deemed desirable or is required
- The schedule of constructing replacement ships.

These factors are interrelated. For a fleet size of 12, it is necessary to deliver one carrier every four years if ships are to be retired at 48 years of age. (Below, we consider the reasons behind this nominal life span.) More generally, at steady state,

$$\text{fleet size} = \text{retirement age (yr)} \div \text{delivery interval (yr)} \qquad (3.1)$$

We display this relationship over build intervals of two to six years and for fleet sizes of 10, 12, and 14 ships, in Figure 3.1. To sustain a 12-ship fleet, vessels can be replaced at age 48 if a new ship is delivered every four years. If the average interval between new ships slips to 4.5 years, then either the retirement age of the ships being replaced must increase to 54 years or the fleet size must be decreased to less than 11 ships. Conversely, if the interval between new ships is decreased to 3.5 years, then either the retirement age can be reduced to 42 years or the fleet size can be increased to greater than 12 ships.

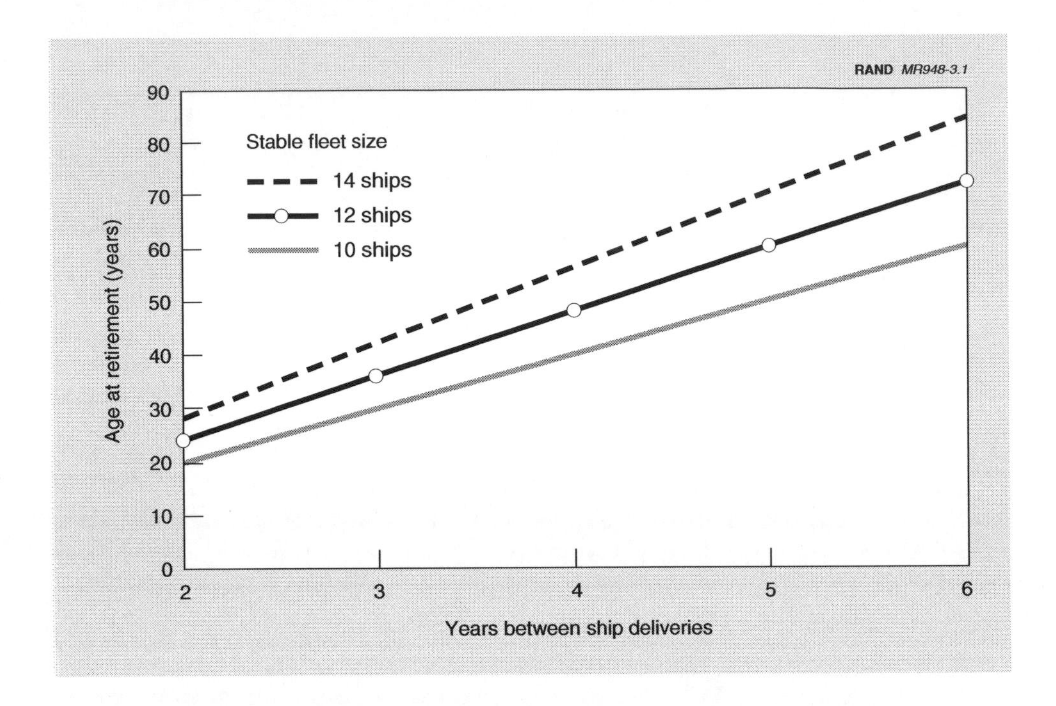

Figure 3.1—Relationship Among Fleet Size, Ship Retirement Age, and Build Rate

Over the long term, this simple 3-variable relationship will dominate the management of fleet composition and the shipbuilding program. And there is good reason for taking the long view: At least a decade is required to budget for, and to construct, a new carrier, and that ship will typically last for nearly half a century of operational service.

Ship-construction decisions covering the next 10 to 20 years cannot be made in isolation from the particulars of current fleet composition. The age distribution of the fleet, at least for the older ships, does not reflect a smooth, stable replacement program. Because of a surge in construction around 1960, three ships are now approaching their 40th year in service. And the demands of national defense and the effects of the national economy on defense budgets have been anything but stable in recent years. Furthermore, different ships have somewhat different capabilities and durabilities, which also influence when ships are retired from the fleet. In the next subsection, we discuss the implications of these factors for the replacement schedule.

Limits on Retirement Age

The range over which retirement age can be reasonably varied is an important question, because, to the extent that there are constraints on retirement age, the only way to achieve a desired fleet size is to sustain a corresponding delivery rate.

In the past, most carriers were retired because they became technologically obsolete. As discussed in Chapter Two, the 1950s and 1960s were marked by a desire for carrier-based operations of aircraft with higher take-off and landing speeds and the benefits of nuclear power. During that period, ships were usually retired after 20 to 25 years (see Figure 2.1). How long a carrier could be made to last was not an issue. Only three carriers were kept for extended periods: *Lexington* (CV 16, 49 years), *Midway* (CV 41, 47 years), and *Coral Sea* (CV 43, 44 years). And of those, *Lexington* was used for training in its later years and was not considered part of the operational fleet.[1] Thus, until quite recently, the Navy had only limited experience in maintaining a carrier for full operational service for more than about 30 years, which was roughly the service life expected when the ships were designed and built.

Beginning in the 1980s, with a production rate of only three new ships per decade, it became impossible to sustain the desired fleet size without extending the operational life of some ships. To date, as shown in Figure 3.2, seven ships (hull numbers 59 through 65) have been extended to an operational life of about 40 years. Yet, unless ships are constructed faster than one every four years, if a 12-ship fleet is to be sustained, carrier life will have to be extended to 45 years or longer.

To achieve the desired longer life, a Service Life Extension Program (SLEP) was performed for CVs 59, 60, 62, 63, and 64 during the 1980s and early 1990s.[2] SLEPs can add about 15 years to the nominal design life of 30 years. For several reasons, this "limit age" should not be interpreted as a precise number: The wearout mechanisms are not well understood, experience with operating older ships at full operational tempo is very limited, and the level of investment in prior maintenance activities has an effect. As the extended nominal life expectancy of 45 years is approached, it might be necessary to perform additional work if extending the ship life still further is desired.[3] However,

---

[1] *Coral Sea* and *Midway* were active until they were retired. However, we were told that they were maintenance nightmares.

[2] CV 67's last complex overhaul was more intensive than usual and had some of the characteristics of a SLEP.

[3] The ship may continue to wear in ways similar to experience to date, or unexpected wearout modes may be encountered. It may not be possible to arrest some modes, and the ship may have to be retired.

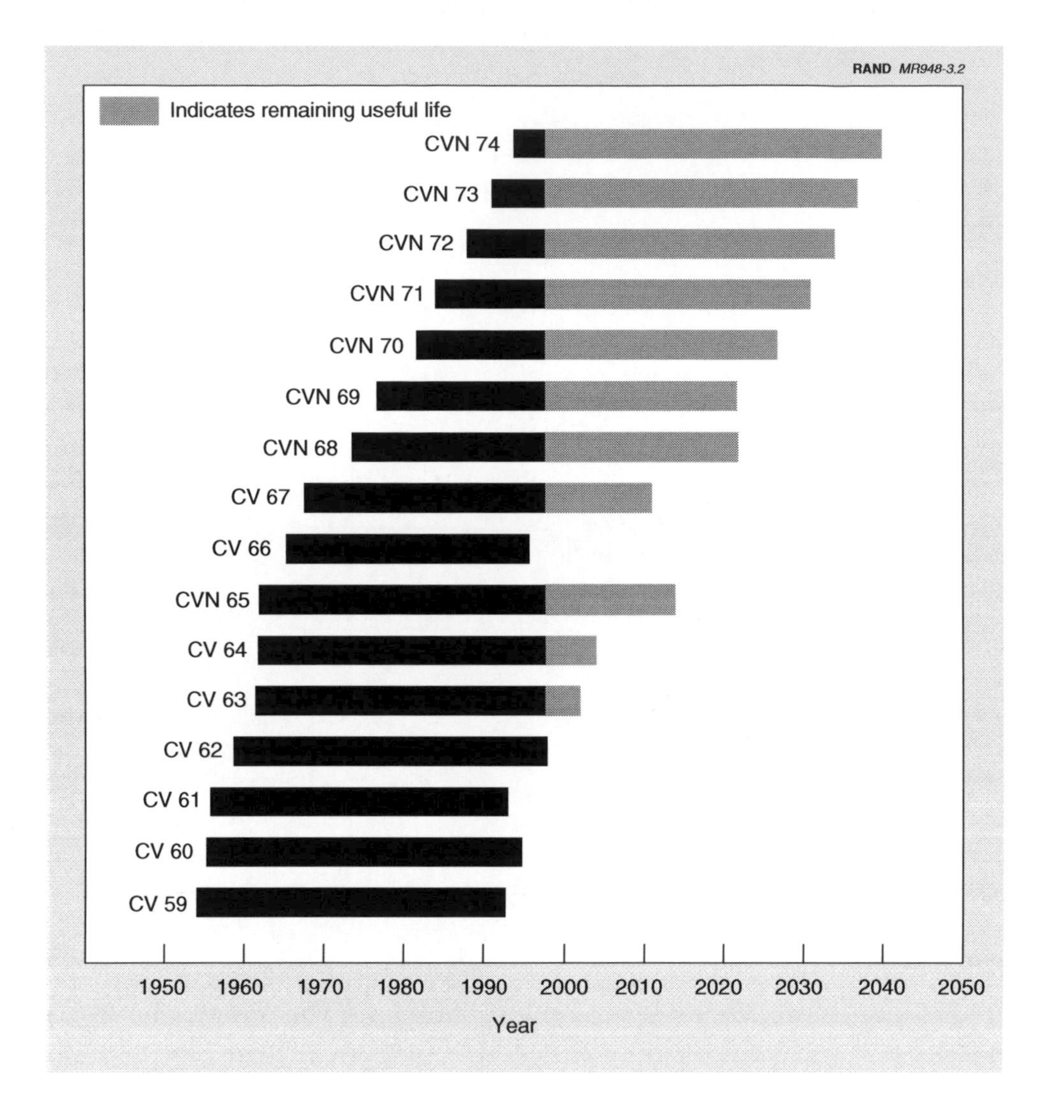

Figure 3.2—Service Lives of Carriers Commissioned Since 1955

continued extension of retirement age should be regarded as increasingly risky, simply because of the lack of experience with operating carriers (or any other class of ships) in such an age range.

For nuclear-powered ships, another factor places practical limits on ship life. At some point, the reactor core becomes depleted and an expensive refueling becomes necessary.[4] To date, only *Enterprise* (CVN 65) has been refueled. Its current fuel supply should permit the ship to operate until late 2013 or early 2014, when it will be 52 years old.

The Nimitz-class carriers' reactor design is different from that of *Enterprise*. In the Nimitz plant, each new set of reactor cores is expected to last about 23 years under normal operating tempo. One refueling has been planned for each ship, at midlife, during a shipyard availability[5] that lasts about two-and-a-half years. A second refueling is not judged a wise or practical expenditure of funds, given hull wearout, the tendency of maintenance costs to rise with age, and the need for more and more extensive backfit to bring aging ships up to the technological state of the art of the newer ones. Therefore, the full operational life of a Nimitz-class carrier can be expected to be 48 to 49 years, assuming that appropriate maintenance on the other parts of the ship is performed when needed.

THE PRODUCTION SCHEDULE OVER THE NEXT 25 YEARS

We now have enough information to permit us to examine a variety of possible ship-construction schedules for the next two to three decades so that we can understand more specifically the effects of ship-delivery rate on fleet size and ship-retirement age. First, however, we need to choose a baseline delivery rate. Whereas, in steady state, we would choose a 4-year interval, the somewhat irregular build schedule that has resulted in the current-fleet age distribution may permit a longer near-term interval. We proceed to analyze whether such an interval is feasible.

Figure 3.3 shows what happens to retirement ages of ships now in the fleet under carrier-construction programs having 4- or 5-year start intervals. We can infer from the figure that past variations in the construction schedule will not permit a near-term sequence of delivery intervals exceeding four years. Specifically, waiting five years between ships means that service lives for *Eisenhower* and *Vinson* will be necessarily longer than the 49 years at which these vessels are expected to exhaust their nuclear fuel supplies.

We now define a nominal schedule, outlined in Table 3.1 and illustrated in Figure 3.4, in which CVN 77 is delivered in 2008 as originally planned and a

[4]The end-of-fuel date might be postponed by reducing operating tempo, but that reduction cuts effective force size by a fraction of a ship.

[5]An *availability* is a period when the ship is scheduled to be in the shipyard for maintenance.

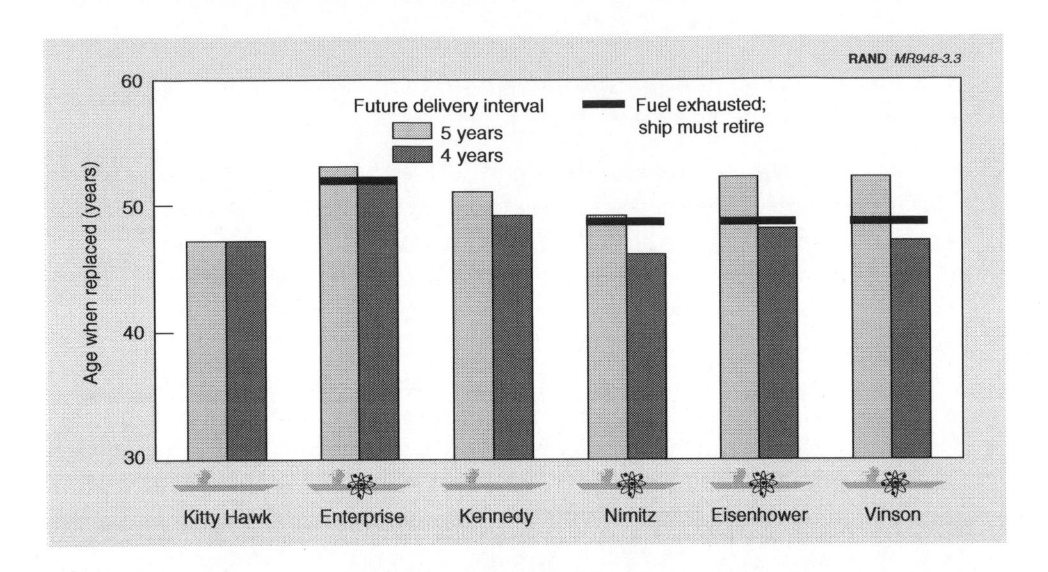

Figure 3.3—Build Intervals for Sustaining a 12-Ship Fleet

Table 3.1

Nominal Ship-Replacement Plan

Ship Replaced	Name	Year Delivered	Nominal End of Life[a]	Replaced By	Year Replaced	Age When Replaced (yr)
CV 64	*Constellation*	1961	2006	CVN 76	2002	41
CV 63	*Kitty Hawk*	1961	2006	CVN 77	2008	47
CVN 65	*Enterprise*	1961	2014	CVX 78	2013	52
CV 67	*John F. Kennedy*	1968	2013	CVX 79	2017	49
CVN 68	*Nimitz*	1975	2024	CVX 80	2021	46
CVN 69	*Eisenhower*	1977	2026	CVX 81	2025	48

[a]The *nominal end of life* is defined as 45 years after delivery for conventional ships and as the end of core life for nuclear ships.

force size of 12 operational ships is sustained. For most ships, we assume the 6.5-year construction period currently planned for CVNs 76 and 77, but we allow an extra year for the two succeeding ships, which will be the first two ships of the CVX class. (The 6.5-year build period for CVN 77 may be somewhat optimistic, given that elements of the industrial base will have to rebuild capacity after the 6-year gap following CVN 76.)

The Navy plans to replace CV 64 with CVN 76, which is scheduled for delivery in 2002. The Navy has not announced plans for retirement of other ships, so we assume that those ships will be retired in order of hull number as new ships are

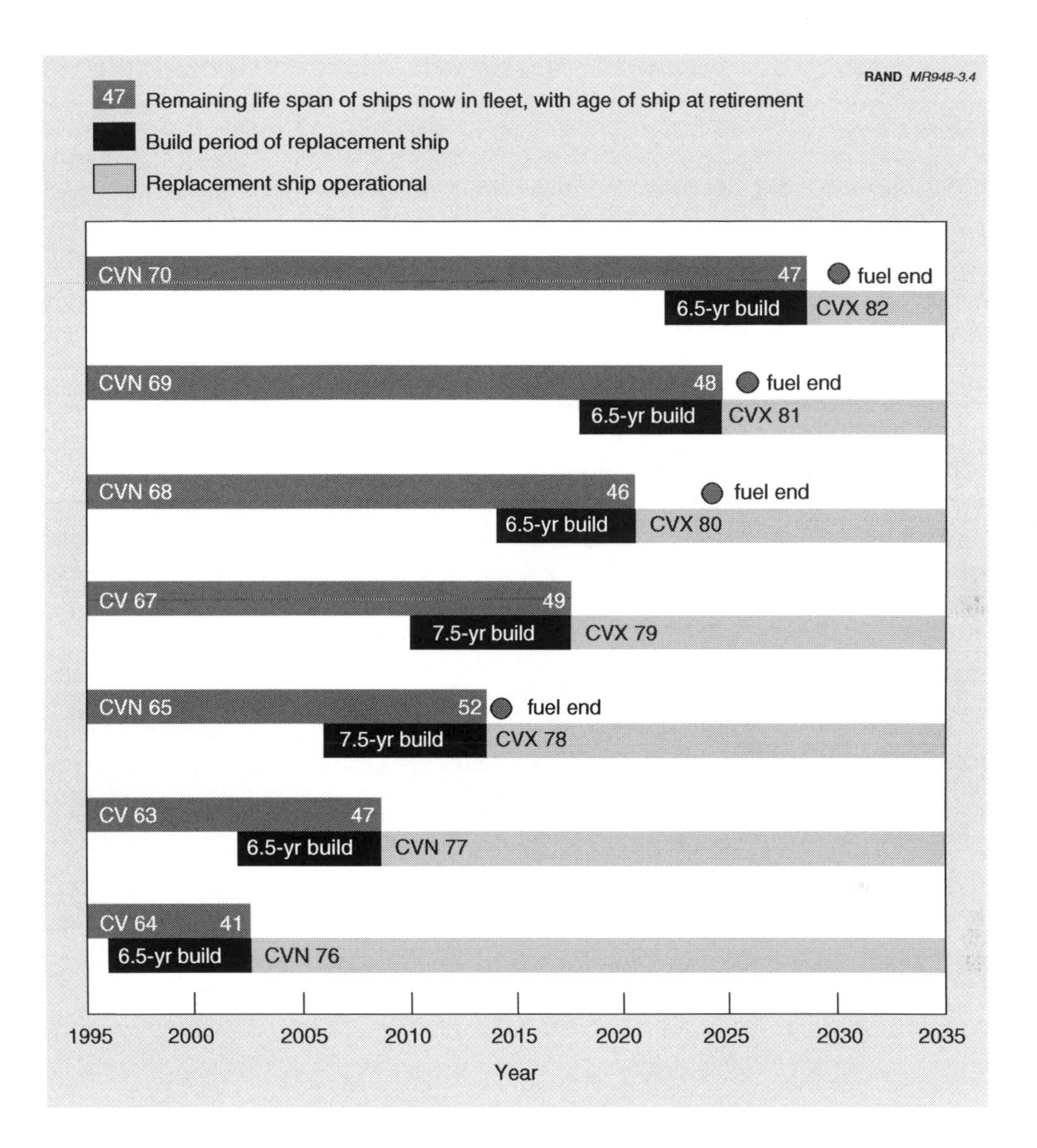

Figure 3.4—End-of-Fuel Dates for Current Carriers in Relation to Nominal Ship-Construction Schedules

delivered. With respect to CV 63 and CVN 65, that order is consistent with the consensus of Navy officials that maximum force capability will be achieved by keeping *Enterprise* in the fleet until it can be replaced by the first CVX design.[6]

---

[6]As mentioned in Chapter Two, after CVN 77, the DoD plans to procure a new aircraft-carrier design, provisionally called the CVX. The design for this ship should be finalized by 2004; CVX 78 is planned to replace *Enterprise* in 2013. Funding for construction of the ship is now planned for FY06.

The plan also assumes delivery of replacement ships on a relatively smooth schedule of one every four years as shown above. The postulated delivery dates for CVN 77 and CVX 78 are as early as those ships are likely to be available.

Two aspects of this nominal plan deserve emphasis. First, it is already too late to achieve a ship-replacement schedule that does not require operating some conventionally powered ships beyond their already-extended 45-year planned service lives. The risks and costs of such operation are unknown, but prudence suggests that they not be treated as trivial. Second, the most critical milestone in the early future is replacement of *Enterprise*. If the present nuclear cores follow the expected depletion rate and are expended in about 2014, then it will be essentially impossible to extend that ship another few years. Even assuming an abbreviated refueling/complex overhaul (RCOH) is performed, the refueling and associated maintenance overhaul would be costly and, even more important, could require a year or more to perform. Thus, it is especially important that delivery of CVN 77 and the first CVX be achieved close to the present schedule.

The constraints imposed by the end-of-fuel dates for the nuclear carriers are indicated in Figure 3.4. A slip in the CVX 78 delivery date will force the fleet size to drop below 12 when *Enterprise* runs out of fuel. If NNS builds CVX, the CVN 77 start date cannot be slipped more than a year or so. Otherwise, near-simultaneous construction demands from the two ships will severely strain, and may exceed, the capacity of the nation's nuclear-carrier construction facilities.

Once *Enterprise* is replaced, there is a little flexibility in the schedule. Note, however, that the assumed delivery dates for CVXs 81 and 82 are within a year or so of the end-of-fuel dates for the ships they are replacing. Thus, CVX 79 and CVX 80 cannot be much delayed, for the same reason CVN 77 cannot be.

Effects of Skipping CVN 77

Some people have suggested that because CVN 77 is the last of a class and because a new design is being planned, it might be desirable to skip construction of CVN 77 and move more quickly to the new CVX design. We have shown that the schedule for replacing some of the older carriers is already tight. Under what conditions might it be practical to skip CVN 77?

If a fleet size of 12 ships is to be maintained, then CVN 65 must be replaced not later than about 2014. Assuming that the first CVX (now CVX 77) replaces CV 63, then CVX 78—now the second ship of its class—must still be finished by 2014. We assume that construction of the first and second CVX ships will require 7.5 years, and we allow a 3-year interval between starts; the result is a start date for the first CVX of about 2004 (and replacement of CV 63 at 51 years of age in 2012). Whether such an accelerated start for the CVX design is practicable cannot yet be assessed, because the basic design parameters have not yet been defined.

Changing Force Size

We have observed that needs and budgets change, and that a force size of 12 might not be right for the long-term future. Thus, we need to explore ways to change the force size—particularly to increase it, which is the more demanding alternative.

Retirement ages are already high under the nominal plan, so force size can be increased only by delivering new ships at shorter intervals. But the long build periods, the need to retire aging ships, and the likely incremental nature of any buildup program limit the rate at which the fleet can be increased. For example, consider the outcome of a near-term attempt to increase force size by two ships, to 14. First, we make the optimistic assumption that the starts of both CVN 77 and CVX 78 are moved ahead two years, to 2000 and 2004, respectively, and that subsequent ships are started every three years thereafter. Second, we assume that existing ships would be retained in the force as long as possible, up to the full 49 years for the nuclear-powered ships. We also assume that only one yard, NNS, is building carriers.

These assumptions yield the time profile of force size in Figure 3.5, which indicates that a fleet size of 14 cannot be sustained until around 2016, even when the plan described above—an aggressive one by current standards—is implemented. That schedule could be accelerated a bit if the delivery rate increases to a frequency even higher than one ship every three years. However, because it typically takes seven to eight years to construct the first ships of a new class, further acceleration would not yield much reduction in the time required to achieve a 14-ship fleet.

Perhaps, instead of sustaining or increasing the carrier fleet, the impetus in the near future will be to reduce it (see Chapter Two for related issues). Whereas force size can be increased only by changing the delivery rate, there are several ways to decrease it. All of them amount to cutting back the delivery rate (e.g., by delaying or canceling the construction of new ships), or retiring ships early, or both.[7]

[7]Neither the options we present here for illustrative purposes nor the order in which we present them should be taken to suggest a preference for any one option. Determining which method is best would require a thorough cost analysis of the various options; such an analysis was not an objective of this research.

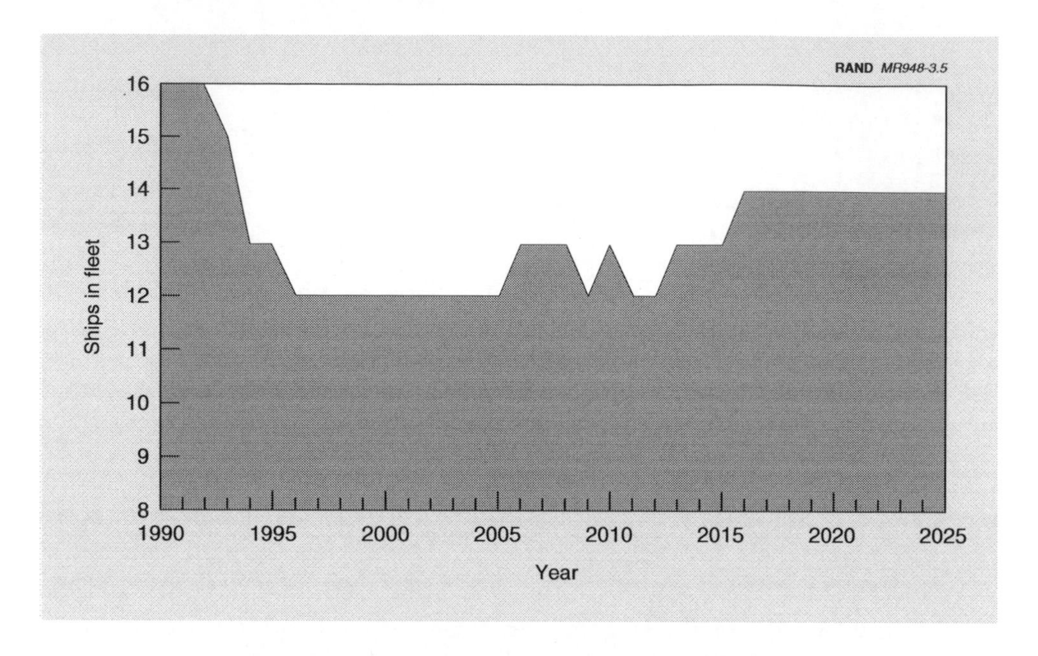

Figure 3.5—Time Required to Increase Fleet Size

For example, suppose a decision is made to reduce the force size to 11 carriers early in the next decade. One option would be to remove CV 63 from the force without replacing it—that is, to skip the construction of CVN 77. Or, it might be preferable to build both CVN 77 and CVX 78, but to allow some slip from the nominal schedule outlined above while still retiring CV 63 and CVN 65 on schedule. Another option would be to build both CVN 77 and CVX 78 on schedule, but to retire CV 63 and CVN 65 somewhat earlier than planned, thus avoiding some operating costs and possibly some maintenance costs on those ships. Each of these options would cause the force size to fall to 11 ships. However, if subsequent construction is sustained at a rate of about one ship every four years, the force could eventually be rebuilt to 12 ships in the future.

CONCLUSION

If force size is to be sustained at 12 ships, two new carriers (beyond CVN 76) must be delivered by 2014. At 52 or 53 years of age, two currently operating ships would then be much older than any fully operational ship has been to date, and one of them—*Enterprise*—will run out of nuclear fuel. Allowing a nominal seven to eight years for construction would mean that the second of those two replacement carriers must be started around 2006. Limitations on ship-construction budgets and carrier-construction facilities make start intervals of less than three years problematic.

Thus, shipyard work on the first of those carriers—CVN 77—must for all practical purposes begin by 2003. The planned start date of 2002 would be consistent with the 4-year construction interval that will sustain a 12-ship fleet when ships retire after 48 years of service—the life expectancy now planned for the Nimitz class.

Chapter Four

HOW THE CVN 77 START DATE AFFECTS SHIPBUILDER COST

We have now defined an approximate date by which CVN 77 must be started if a 12-ship carrier fleet is to be sustained. Our next task is to determine how CVN 77 construction cost is affected by varying the start date from the earliest feasible date to the latest date. Considering that design work on CVN 77 remains to be done, we take the earliest feasible start date to be 1999, and we have shown the latest start date to be 2003. (To understand the results better, we also look at start dates somewhat later than 2003.)

We expect some cost variation because Newport News Shipbuilding constructs and maintains other vessels, and costs associated with acquiring workers for carrier construction should vary as labor demand for those other projects rises or falls. However, while we expect and account for extra costs associated with new hires, we do not expect or account for any extra costs associated with the passage of time since construction of the last carrier. In our analysis of submarine production, such costs were important.[1] However, in the current analysis, we found no skills required for carrier construction that are not also needed for other construction and maintenance projects scheduled for Newport News (see box, next page).

THE SHIPYARD WORKFORCE COST MODEL

Construction of a Nimitz-class nuclear aircraft carrier is a very complex and costly process. The actions of thousands of workers in various skills and trades must be efficiently and effectively orchestrated over six to eight years. Materials will have to be in the right place at the right time. And, at Newport News Shipbuilding (NNS), the demands for skilled labor for carrier construction must

---

[1]See John Birkler et al., *The U.S. Submarine Production Base: An Analysis of Cost, Schedule, and Risk for Selected Force Structures,* Santa Monica, Calif.: RAND, MR-456-OSD, 1994a. Although the fact that work on other types of ships does little to preserve submarine-specific skills was important in this study, the most significant consideration was reducing a shipyard's workforce to a small cadre, then re-expanding it.

be integrated with similar demands from various other construction and maintenance projects. To help us understand the cost implications of varying the start date and construction period of CVN 77, we built an analytical model of the requirements for and availability of skilled labor at NNS.[2]

Industrial Base Adequate to Support Further Carrier Construction

On the basis of interviews with shipyard, vendor, and Navy personnel and after analyzing the data they have provided, we conclude the following: The Newport News Shipbuilding facilities, and the supporting industrial base throughout the United States, are expected to retain the basic capabilities necessary to build large, nuclear-powered aircraft carriers into the foreseeable future, regardless of when or even whether CVN 77 is built. However, failure to start CVN 77 in the 2000–2002 time frame will inevitably lead to some decay in the quality of those capabilities and, hence, to increased costs, schedule durations, and risks when the next carrier is started. While other work may employ similar skills, the current and projected workload does not maintain the volume of skills to build CVN 77 if the ship is delayed and would require significant reconstitution cost for both shipbuilding skills and selected component suppliers.

The reconstitution required would add to the cost of the next carrier and the time to build it. And the cost of other Navy work in the yard would increase as fixed overhead costs were spread over a smaller volume of work.

Assuming CVN 77 is built, its start date could result in cost increases or decreases related to industrial-base efficiencies (as detailed in this chapter). We believe that, if CVN 77 is started in 2003, it will be no more of a challenge to build than if it is started in 2000.[3] There will be no carrier-specific shipyard skills to deteriorate in that interval. Furthermore, workers may be transferred from other projects to carrier construction without having to go through special training and incurring the additional costs such training entails.

[2]In building this model, we drew on the RAND team's experience with previous industrial-base analyses, e.g., Birkler et al., 1994a; Birkler et al., *The U.S. Submarine Production Base: An Analysis of Cost, Schedule, and Risk for Selected Force Structures: Executive Summary*, Santa Monica, Calif.: RAND, MR-456/1-OSD, 1994b; John Birkler et al., *Preliminary Analysis of Industrial-Base Issues and Implications for Future Bomber Design and Production*, Santa Monica, Calif.: RAND, MR-628.0-AF, October 1995; and John Birkler et al., *Reconstituting a Production Capability: Past Experience, Restart Criteria, and Suggested Policies*, Santa Monica, Calif.: RAND, MR-273-ACQ, 1993.

[3]This is not to say that a construction effort started in 2003 will go smoothly. As in any large, complex construction project, a few components may not fit or work properly when first installed. These will have to be redesigned and modified.

Our model uses a linear programming formulation to develop a least-cost solution for matching the demand for and supply of skilled workers across all the projects scheduled at NNS. The model calculates the net present value[4] (expressed in FY98 dollars) of the future costs of hiring, training, and terminating workers of various skills as it adjusts the workforce to meet current and future demands for labor. Increases in the workforce are constrained by the time needed to acquire and train personnel to various proficiency levels. They are also limited by the personnel available for mentoring trainees. The model accounts for differences in efficiency and compensation between trainees and experienced workers.

Managing the labor force at a shipyard as diverse as NNS is a complex problem.[5] Skilled shipyard and project managers use various techniques to balance and sequence the demands for various skills. The model cannot capture all of these techniques, but it does account for the use of overtime and the hiring of temporary workers (through outsourcing and subcontracting), ways in which additional labor is made available to meet peak demands.

No model can ever represent the full range of options in such a large, complex process. A trade-off must be made between the "granularity," or detail, of the model and the speed with which the model produces solutions. As more detail is added, the model's complexity grows exponentially. The level of detail currently represented in the model reflects both the level of detail of the shipyard data available to us[6] and our desire to produce solutions fairly quickly. The model should be viewed as a planning tool, designed to show the general implications and trends of various production strategies. The model can very quickly narrow a wide range of options to a much smaller set that is amenable to further examination. Its accuracy is not sufficient for detailed budgeting or cost-estimating exercises, however.

In the remainder of this section, we describe how the model represents the demand for skilled labor and the supply of such labor, and how the model identifies workforce levels that cost the least over time.

---

[4]*Net present value* is the sum of the discounted values of a time series of outlays and returns. It is defined at greater length in Chapter Seven.

[5]Newport News Shipbuilding currently employs 18,000 persons; 60 percent of this workforce is covered by collective labor agreements. In 1995, the company and the United Steelworkers of America (USWA) reached a new labor agreement, which extends to April 1999. Under the agreement, wages have been frozen and new job classifications are offered that give more flexibility to move persons between projects.

[6]The design of the model allows more detail to be incorporated to address specific questions of interest. However, additional data would have to be made available.

Demand for Skilled Labor

The model starts with a list of current and future projects scheduled for NNS. Each project in the list has an initially specified start date and completion date.[7] For each project, the model takes as input the number of workers required, by quarter, for each of nine aggregate skill groups.[8] Different skill-demand profiles can be specified for different build periods, or the model will proportionally scale the workload demands for shorter or longer build periods. As an example, the currently planned 6.5-year construction profile for CVN 77 is shown in Figure 4.1.[9] The demand profiles for each project can be combined to produce the overall demand by skill for NNS. An example of this total shipyard labor demand, by skill group, is shown in Figure 4.2 (at smaller scale than in Figure 4.1).

The total-labor-demand profile can also be segregated by project as shown in Figure 4.3. Here, all work under contract as of the beginning of FY96—including the ongoing construction of CVNs 75 and 76 and other projects scheduled to run out by 2003—is aggregated under the "NNS" label. The remainder of that band represents a fairly constant, long-term allowance for engineering-development work. The other bands show the anticipated labor demand for NSSN construction and for availabilities of various carriers, including all availabilities for CVN 65 and the midlife RCOHs of the Nimitz-class ships.[10] Note that the sequence of RCOHs provides a substantial, long-term demand, albeit a cyclical one.

Figures 4.2 and 4.3 do not include the labor demand for building CVX 78. Labor for that ship is included, as appropriate, in the results for the analyses discussed later in this chapter, "Taking CVX into Account." However, the rest of this labor-demand profile remains the same across all of our analyses (except for the minor cost-minimizing shifts mentioned above). Note that the profile contains no projection of non-Navy work after the next several years and counts

---

[7]The model captures all of the major Navy and commercial projects that are ongoing or scheduled for NNS. These include remaining construction of CVN 75 and CVN 76, Double Eagle-class product tankers, and other work currently "on the books"; the planned future construction of new attack submarines (NSSNs) and CVN 77; availabilities for *Enterprise;* and refueling overhauls for Nimitz-class carriers.

[8]Currently included in the model are the following skill groups: welding, fitting and fabrication, pipe fitting, electrical, outfitting, machinery, construction support, engineering, and other support. These skill groups represent the organization of NNS and the level of data it provided.

[9]The profile includes about a year of engineering and other support activities occurring prior to construction start, making the total width of the profile 7.5 years. (A few quarters of even lower activity levels during the engineering phase have been omitted.)

[10]We took the CVN 70 RCOH as the model for those to follow. Note that the work profile for RCOHs lasts longer than the 32 months a carrier is scheduled to be in the yard, because some engineering and planning work is undertaken prior to the ship's arrival.

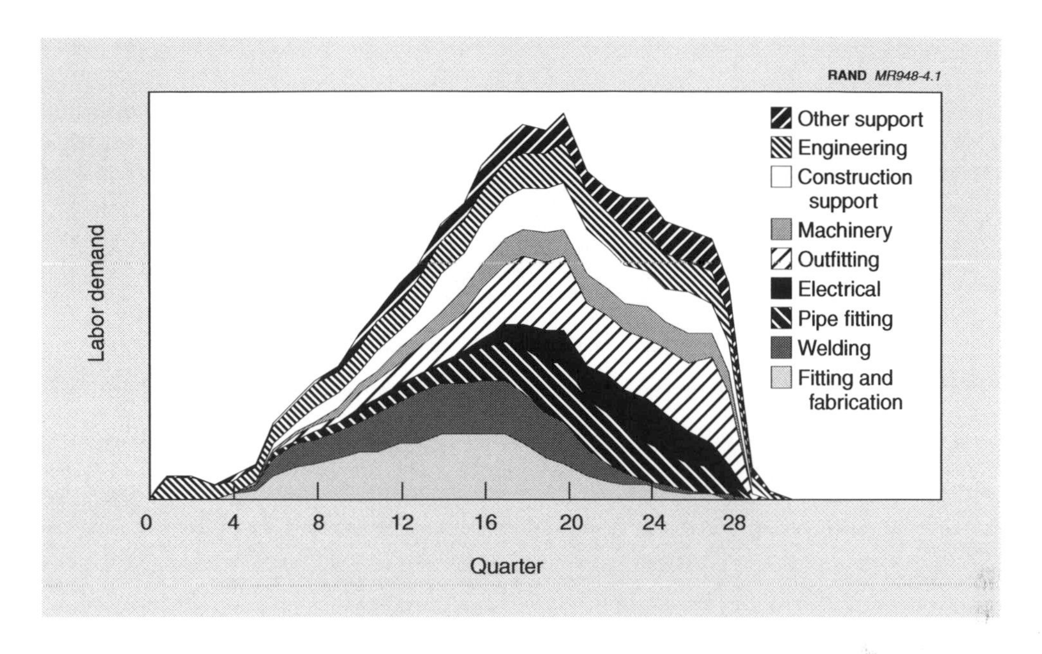

Figure 4.1—CVN 77 Labor-Demand Profile

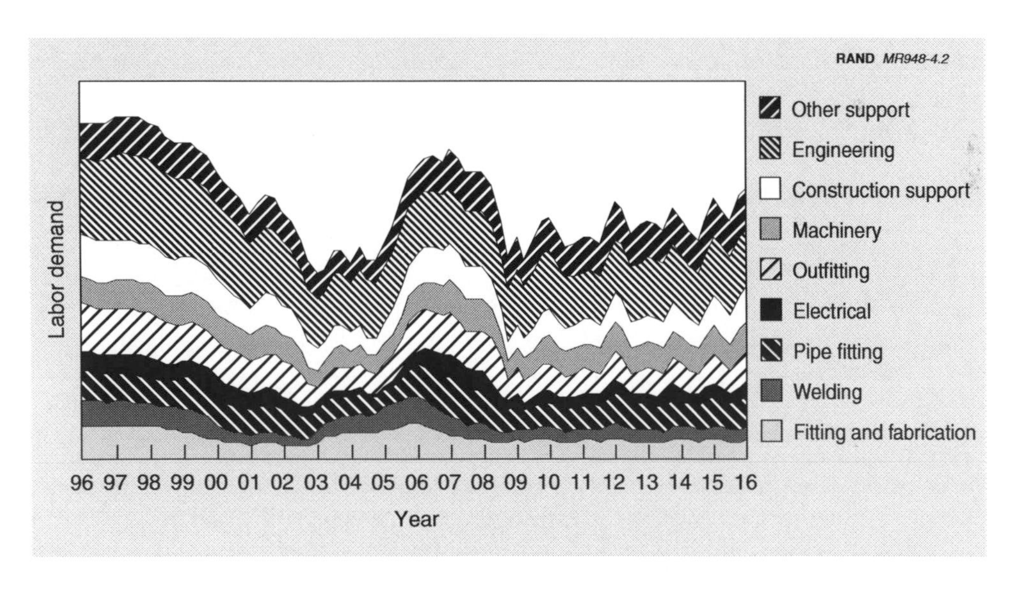

Figure 4.2—Total NNS Labor Demand, by Skill Group, Assuming CVN 77 Starts in 2002

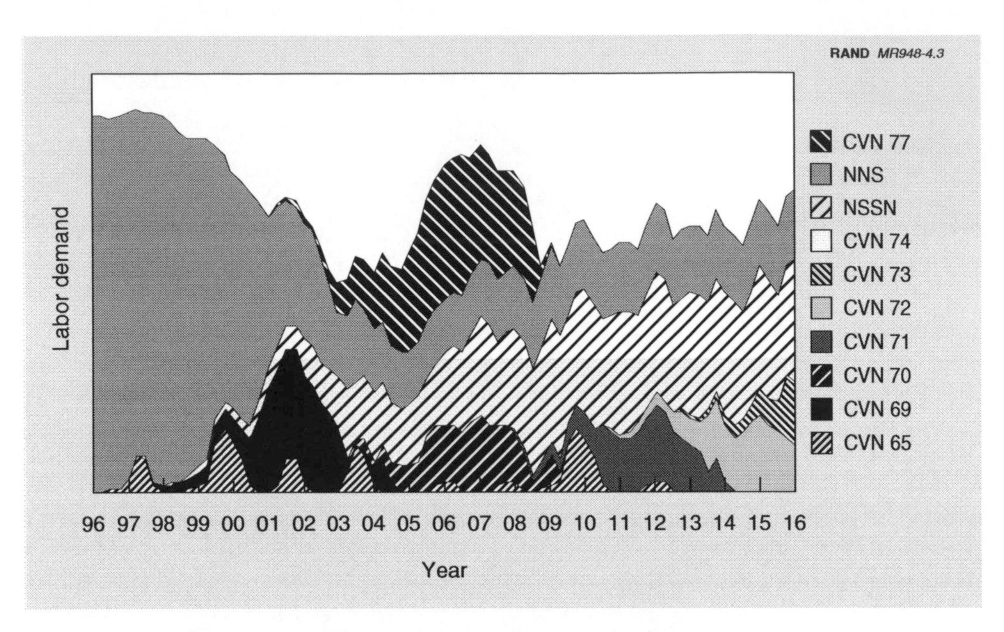

Figure 4.3—Total NNS Labor Demand, by Project, Assuming CVN 77 Starts in 2002

no future Navy projects that NNS may compete for (e.g., Surface Combatant 21) whose size, nature, and builder are now uncertain.

Besides allowing breakdowns by skill and project, the model can provide time-phased requirements for individual skill groupings. The total demand for welding skills is shown in Figure 4.4; the demand due to CVN 77 is highlighted. The various demand profiles by skill and project have proven to be a product of the model useful for understanding current and future shipyard demands. Figure 4.4, for example, clearly shows the swings in the demand for welders resulting from the end of CVN 76 steel fabrication in 1999 and the beginning of major fabrication for CVN 77 in 2003. (The RCOHs shown in Figure 4.3 do not provide enough work for welders to damp out this demand swing.)

Supply of Skilled Labor

The model requires that the supply of each skill group be sufficient to meet the demand.[11] The model does not permit cross-leveling between skill groups, e.g., the use of welders to fit pipes when there is a shortage of pipe fitters and an excess of welders. Thus, labor supply for each skill is modeled separately (as is demand). We combined the results of the nine submodels by skill to give the total labor supply for each quarter at NNS.

---

[11]In addition to labor, equipment and facility supply and demands can also be included in the model, but have not been included for this study.

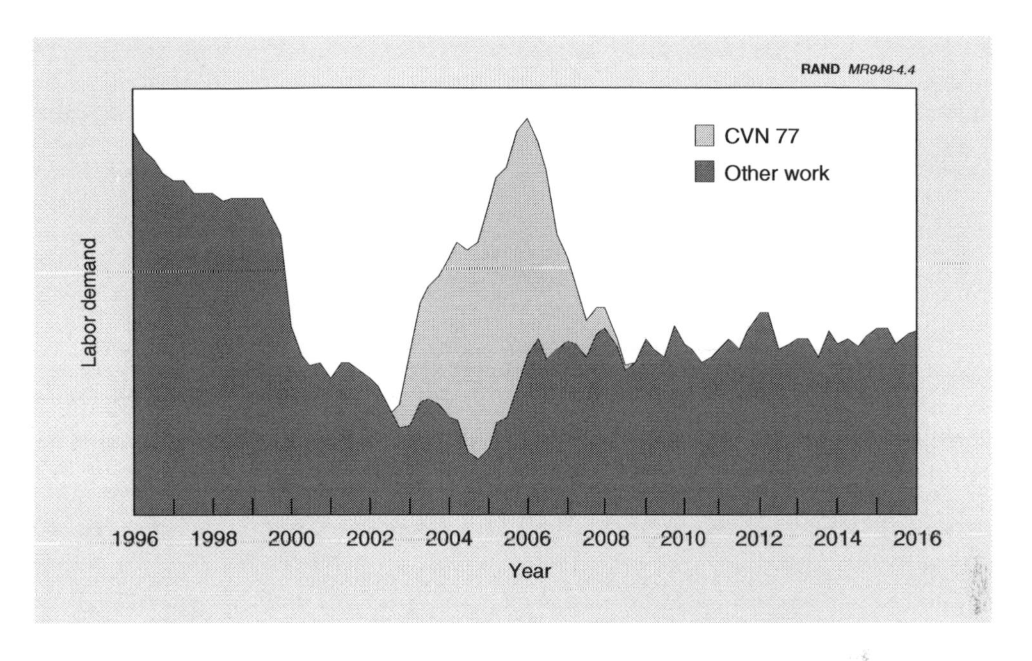

Figure 4.4—Total Demand for Welders at NNS, Assuming CVN 77 Starts in 2002

The model accounts for several categories of labor supply: current shipyard employees working either regular time or overtime, new hires, and temporary subcontracted or outsourced labor. Labor supply within these categories and within skill groups is governed by the following:

- Wage rates that vary by experience level
- Maximum allowable overtime
- Personnel attrition, by experience level
- Cost of hiring employees and the maximum number that can be hired each quarter
- Time and cost to train new employees
- Mentoring, or training supervision, requirements
- Cost and availability of temporary labor.

Also, each skill group has a skill-improvement curve that shows the increase in productivity as a function of time. These curves start new hires at low productivity and, as a result of training, advance an employee through successively higher proficiency levels.[12] The curves have a typical *S* shape, indicating small gains in productivity during early phases of training, followed by more-rapid increases once fully trained, with productivity leveling off after a specified time on the job.

Minimizing Total Shipyard Costs

The model determines the least-cost way to match supply to demand, given the availability of labor and the constraints on hiring and training new employees.[13] Those constraints restrict how quickly NNS can increase supply within a skill group. Therefore, *the model sometimes produces excess supply to ensure that sufficient labor is available to meet future demands.* This excess capacity typically occurs just prior to large increases, or peaks, in workload demand.

An example of such excess capacity is shown in Figure 4.5, where the lower curve is the shape of the welder-demand profile from Figure 4.4 and the dark gray area represents the excess capacity necessary to meet future demands. Here, if the schedule is to be maintained and the peak welder demand in the year 2006 is to be met, hiring must begin in 2001, and the least-cost hiring profile is that shown in the graph.[14] By summing across all the individual skill groups, the model develops the employment profile for the entire shipyard (shown in Figure 4.6).

USING THE MODEL TO ESTIMATE COSTS FOR DIFFERENT SCENARIOS

We use this model of shipyard workforce cost to answer our research question of how various construction start and completion dates for CVN 77 affect the cost of building the ship. Currently in the Navy shipbuilding budget, CVN 77 is authorized for a 2002 construction start and for delivery in 2008. This schedule serves as our baseline. For this schedule and others, the model generates

---

[12]The model does not recognize rehires. All workers in a particular skill group not in the workforce the previous quarter start at a low production level and advance at the same rate—a simplification that overestimates the cost of workforce swings. However, we were informed by NNS personnel that the probability of rehiring a worker laid off a couple quarters earlier is very low, so the overestimate is probably not substantial.

[13]The model minimizes the net present value of future costs. However, because budget projections for ship construction are typically not discounted, we present our results in this chapter in undiscounted terms.

[14]Whether the shipbuilder would actually carry the excess labor (and whether the government would pay for it) is unclear. The shipbuilder may prefer slower workforce reconstitution, coupled with a longer schedule for CVN 77; or it may be able to reschedule other work to help level the workload.

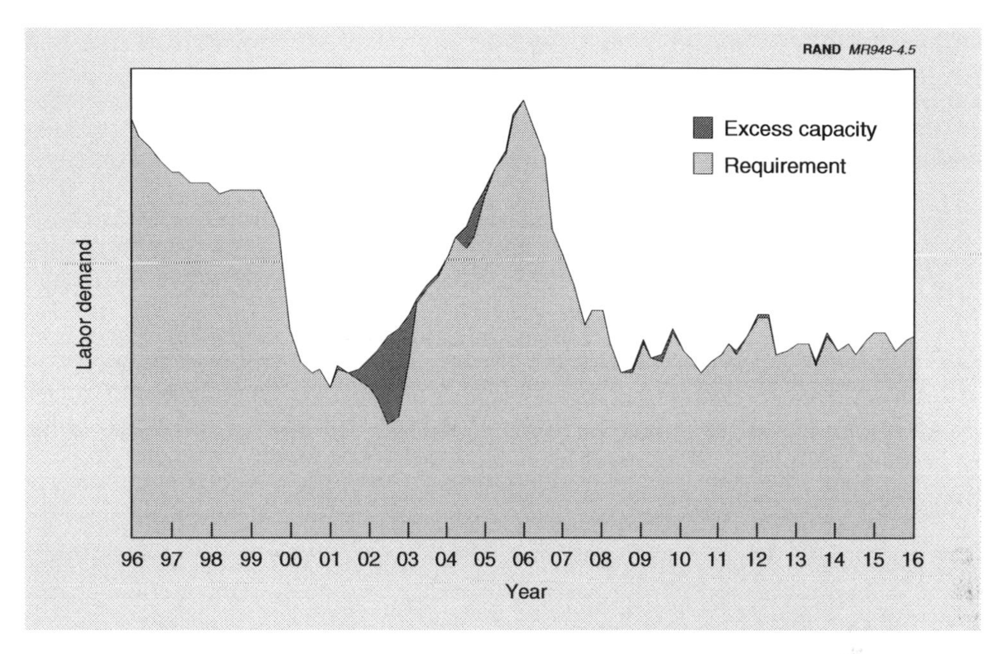

Figure 4.5—Need for Excess Supply of Welders, Assuming CVN 77 Starts in 2002

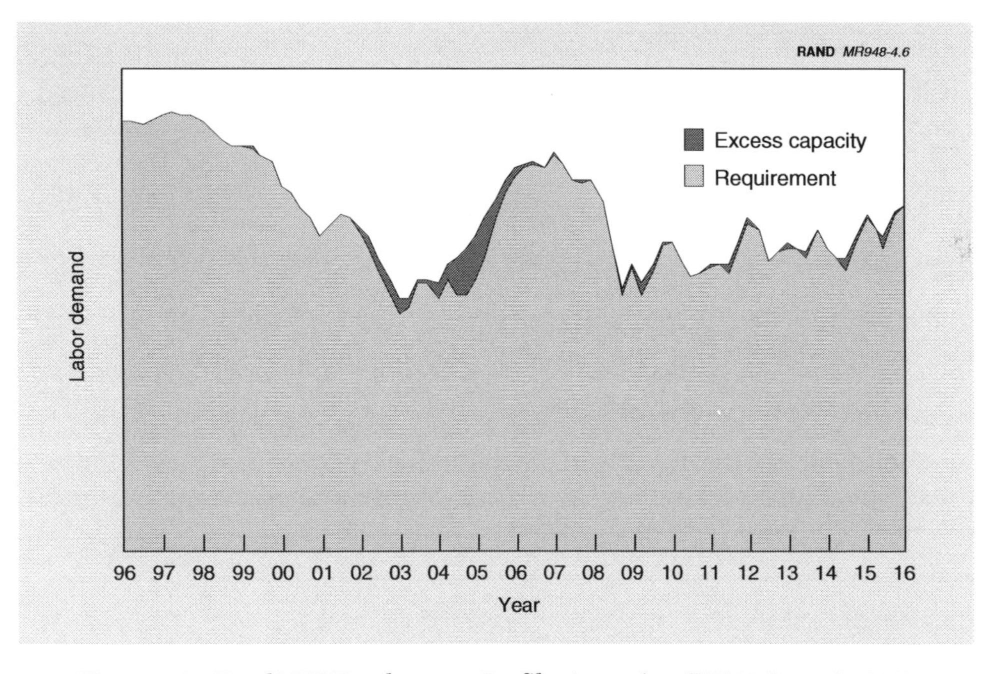

Figure 4.6—Total NNS Employment Profile, Assuming CVN 77 Starts in 2002

shipyard labor costs, to which we add the costs of contractor-furnished equipment (CFE)[15] and shipyard overhead.

Varying the Start Date and Maintaining Build Period

In Figure 4.7, we answer a simple formulation of the research question: How much more or less will CVN 77 cost if its 6.5-year build period is started later or earlier than 2002? Each point on the graph represents the cost of starting CVN 77 in the corresponding year, minus the cost of starting it in 2002.[16] As discussed above, these costs are always the least that can be managed for a given start date and build period. The graph shows that as the start date is delayed, building CVN 77 costs more—over $300 million more—for a delay of just one year. If the ship is started earlier than 2002, money is saved—about a quarter of a billion dollars for a start in 2000.

The analysis for which results are shown in Figure 4.7 did not include CVX as one of the other projects in the yard. That ship could not be started practicably until after 2003 and thus outside the period of policy interest for CVN 77. (Recall from Chapter Three that CVN 77 starts after 2003 will not sustain the carrier fleet at 12 ships.)

Why do we get these results? Whereas the numbers in Figure 4.7 represent all costs to the shipbuilder, labor dominates the year-to-year differences. Over the period shown in that figure, neither CFE costs nor overhead costs vary by more than $100 million from the baseline. And these two sources of variation do not reinforce each other: CFE costs increase to the baseline in 2002, then stay there; overhead costs, at the baseline through 2002, decrease thereafter. (See Appendix D for depictions of cost differences by source.)

The larger variations in labor costs can be understood with the help of Figure 4.8, which shows the anticipated labor level at Newport News over the period of concern. The lower portion of the graph represents the labor level without CVN 77. It drops almost steadily to 2005, then rises. The small, free-floating element in the figure is the planned labor profile for CVN 77. It is positioned to

[15]For convenience, we use the term *contractor-furnished equipment* to refer to all materials bought by the shipbuilder from nongovernment sources. We thus mean to include not only the complex, expensive items typically denoted by "CFE" but also simpler, less expensive purchased parts and even raw materials.

[16]More precisely, each point represents the total shipyard costs if CVN 77 is started in that year, minus total shipyard costs if CVN 77 is started in 2002. We ascribe extra shipyard costs or savings associated with rescheduling CVN 77 to that ship and speak of them as "CVN 77 costs."

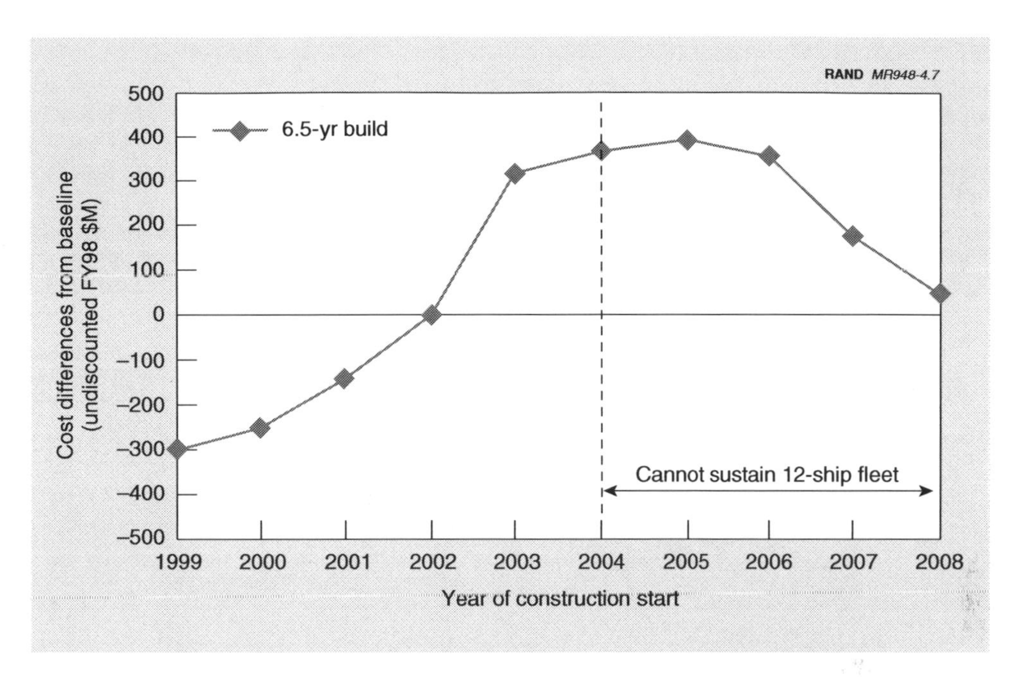

Figure 4.7—Effect of CVN 77 Start Date on Construction Costs

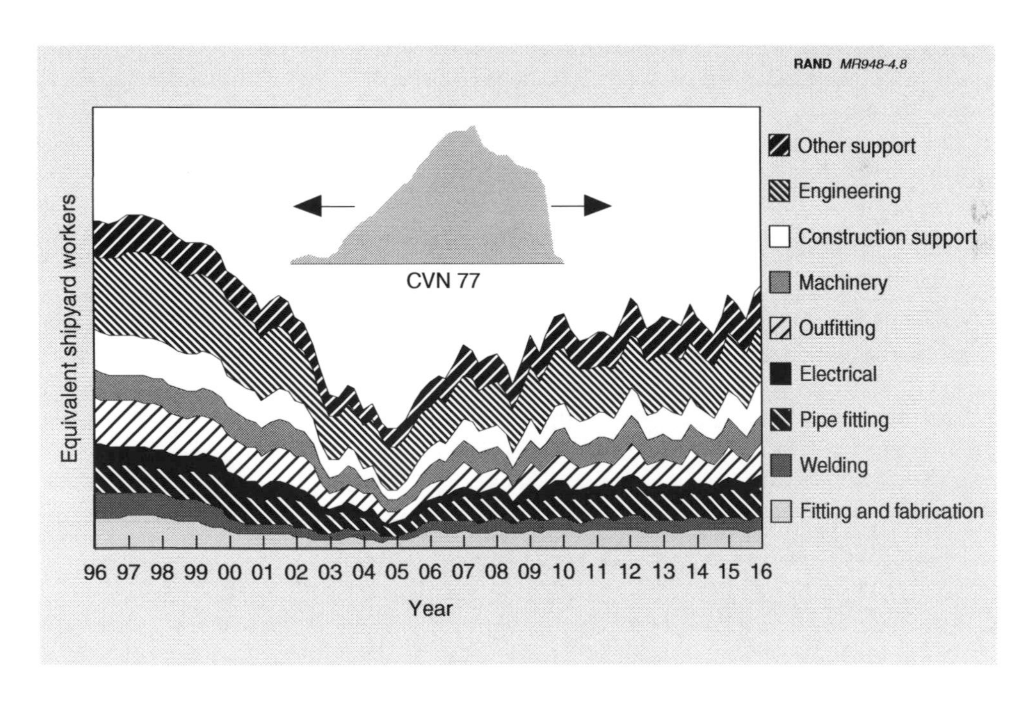

Figure 4.8—Anticipated Labor Level at Newport News over the Period of Concern

correspond to a construction start in 2002. Most of the labor for CVN 77 occurs a little after the valley in the labor-level curve. If the CVN 77 work is moved to the right, it begins to pile onto even higher labor levels, leaving the valley even emptier and requiring greater swings in the size of the workforce. And those swings require outlays of money for rehiring, training, and other workforce inefficiencies. But if CVN 77 is started earlier, the peak labor workload for CVN 77 falls more neatly into the valley and the yard's total workforce requirements become more uniform over time.

Incidentally, the labor-level valley shown in Figure 4.8 cannot easily be filled with commercial work. Figure 4.9 repeats the CVN 77 demand profile with that for a Double Eagle-class product tanker, for comparison. Demand profiles for various shipyard projects are shown in Appendix E.

Varying Both Start Date and Build Period

So far, we have restricted ourselves to a 6.5-year build period for CVN 77. What if more time is allowed? In Figure 4.10, we repeat the curve from Figure 4.7 and add curves representing build periods one year longer and two years longer. As the graph shows, for most start dates of interest, stretching the schedule tends to reduce costs. In fact, for the 8.5-year case, delay beyond 2002 costs very little—but, then, CVN 77 would not be available until at least 2011, when the ship it is to replace will be 50 years old.

It is thus of some importance to choose one desirable delivery date—the planned date of 2008, for example—and examine what happens to costs as build period is lengthened, i.e., as start date is moved earlier. Again, our baseline provides for a 6.5-year build period, meaning a start date of 2002 (represented in Figure 4.10 by the diamond at that date) for delivery in 2008. If CVN 77 is started a year earlier than 2002 and takes a year longer to build than the 6.5 years we have been assuming, the Navy saves almost $300 million (the square at 2001). If CVN 77 is started two years earlier and takes two years longer, the Navy saves close to $400 million (the triangle at 2000). Thus, for CVN 77, longer schedules mean lower costs.

Some readers may find this result counterintuitive, given long experience of seeing acquisition-cost increases correlated with schedule slips. Our result does not violate that historical pattern, which is based on programs that were planned, staffed, and had facilities for one schedule, and then were carried out on a schedule that was lengthened well after the program started and all the planned resources were in place. Some such programs have taken longer because the total amount of work necessary to complete them was under-

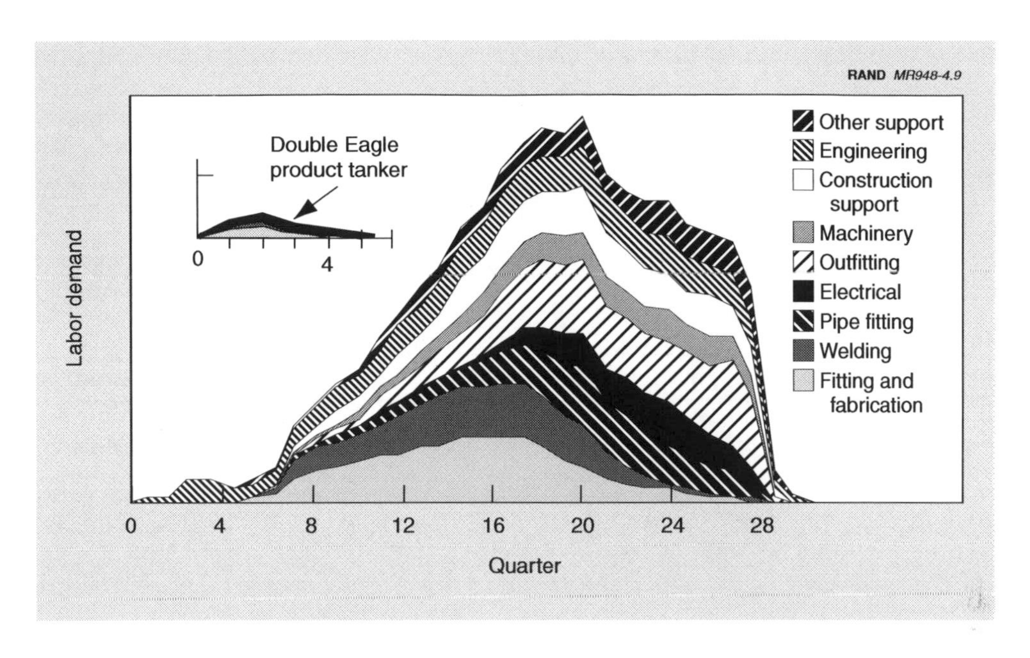

Figure 4.9—Labor Demand of Commercial Projects in Relation to That Required to Build CVN 77

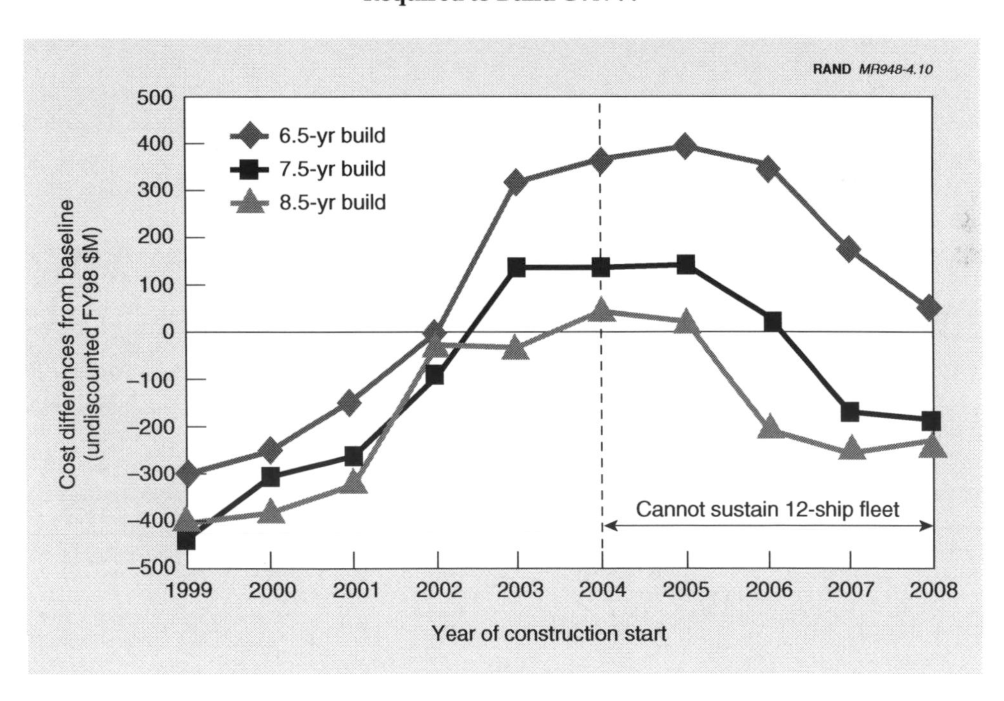

Figure 4.10—Build Periods and Construction Costs

estimated (sometimes because of requirements introduced after program initiation).

In the present case, we are simply changing the *planned* project duration, not the amount of work involved. Therefore, a longer build schedule need not cost more.

Why does it cost less? For one thing, the planned baseline project duration of 6.5 years is considerably shorter than normal for a Nimitz-class carrier. From Figure 4.11,[17] we can see that the average build period, from contract to delivery, has been about eight years, and only one ship (CVN 71) has been built in less than seven years. Information provided by Newport News indicates that a construction time of about eight years generally leads to a more efficient use of staff and facilities. Management obtains more flexibility to schedule tasks to take advantage of month-to-month workforce fluctuations on other projects. Such scheduling flexibilities require higher-resolution data than we have for our model, but NNS estimated a total cost savings of 5 percent for the 8.5-year schedule compared with the 6.5-year schedule. Therefore, we input in the

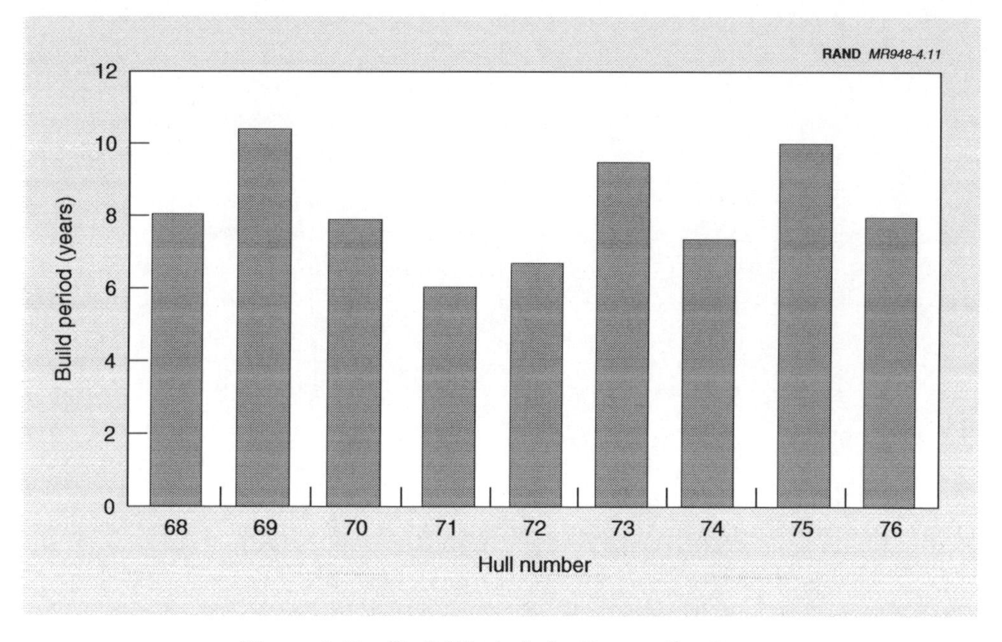

Figure 4.11—Build Periods for Recent Carriers

---

[17]When two ships are contracted simultaneously, the second of the pair generally takes longer to build. Thus, the times for CVNs 69, 73, and 75, all second ships of a pair, are longer than would be expected for CVN 77.

model a total CVN labor demand for the 8.5-year schedule that is 5 percent lower than what we used for the 6.5-year schedule.[18] For a 7.5-year schedule, we interpolate a 2.5-percent savings.

Another reason longer build periods cost less is that, even if there is no flexibility premium, the peak labor demand would be lower. As a result, the costs associated with the quarter-to-quarter workforce fluctuations that our model does account for would also be lower. Of course, the specific savings shown in Figure 4.10 depend further on the way these slightly smaller, flatter labor demands fall out against those for the other projects in the shipyard. In fact, the labor demands of other projects are the sole cause of variation in costs by start year for a given build period, as Figure 4.12 illustrates. As the figure indicates, an 8.5-year build for CVN 77 levels labor demand more than do the other two build periods.

COMBINING SHIPYARD AND VENDOR COSTS

We have demonstrated the potential for savings if shipyard work on CVN 77 is started earlier than 2002. As explained in Chapter Six, the cost of starting in any given year can be lowered further if contractor-furnished equipment is purchased ahead of the shipyard start date. Table 4.1 shows the savings achievable from earlier shipyard start dates and from earlier CFE procurement dates, assuming delivery of CVN 77 in 2008. All the numbers shown are savings relative to starting both CVN 77 and the procurement of CFE for that ship in 2002, represented by the zero in the lower right corner. Then, as was already shown in Figure 4.10, if the ship is started one year earlier (7.5-year build), the Navy saves $260 million; two years earlier (8.5-year build), and the savings is $390 million. But these numbers assume CFE procurement begins the same year the ship starts. If, for any one of these CVN start dates—2000, 2001, or 2002—CFE procurement begins ahead of time, more money is saved, as can be seen by following the columns in Table 4.1 upward.[19] The additional savings are $10 to $20 million if CFE procurement begins a year ahead of the start date for CVN 77, $30 to $50 million if it begins two years ahead.

[18]NNS's lower demand estimate is 5 percent less in the aggregate, but varies among the nine skill groups.

[19]For example, for a 2000 start, CFE procurement in 1999 saves $410M – $390M, or $20M; CFE procurement in 1998 for the same start date saves $440M – $390M, or $50M.

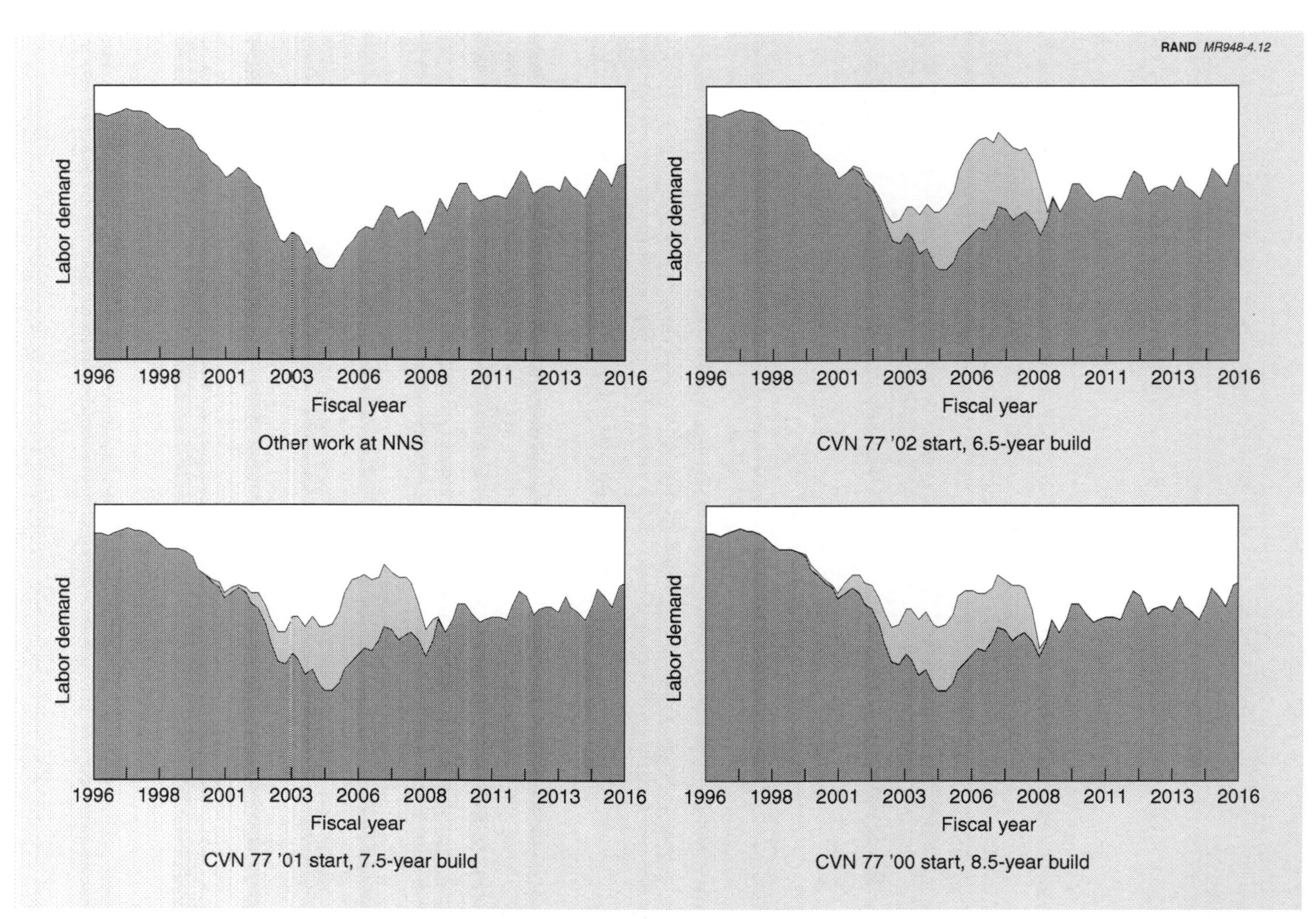

Figure 4.12—Labor-Demand Requirements for Construction of CVN 77 for Different Start Dates and Build Periods, in Relation to Other Work at NNS

Table 4.1

Savings from Earlier Procurement of Contractor-Furnished Equipment

CFE Procurement Start	Savings Relative to Baseline (FY98 $M) for CVN 77 Delivery in 2008 and Start in 2000	2001	2002
1998	440	330	80
1999	410	300	50
2000	390	280	30
2001	—[a]	260	10
2002	—[a]	—[a]	0

[a]Does not correspond to "early" CFE procurement.

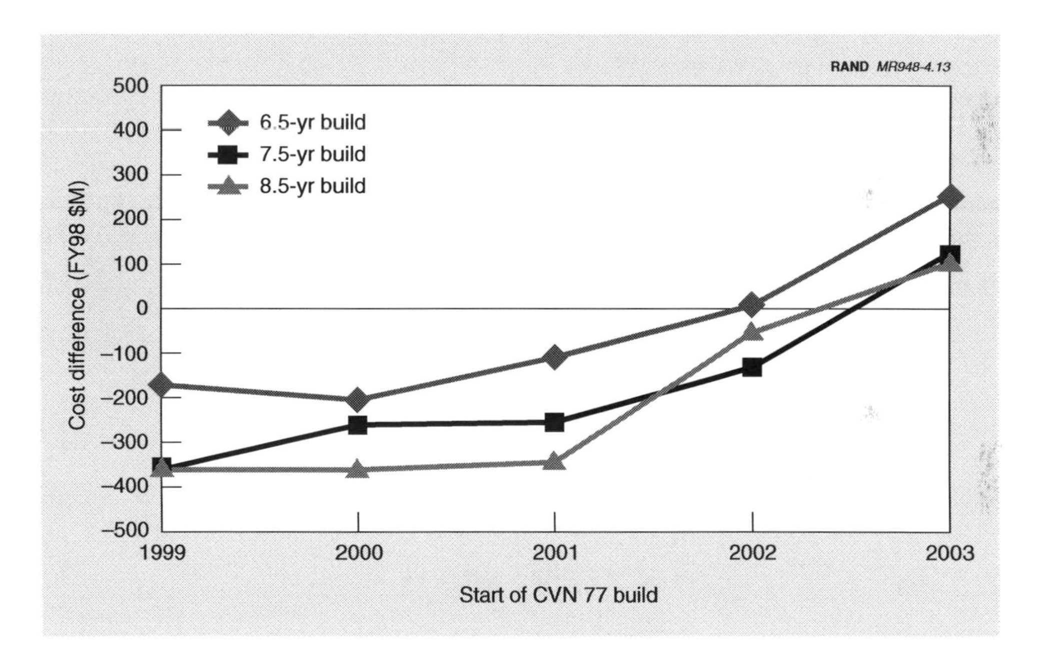

Figure 4.13—Effect of CVN 77 Start Date and Build Period on Construction Costs, Assuming CVX Starts in 2006

TAKING CVX INTO ACCOUNT

The analysis presented above ignores construction of CVX at NNS. In Figure 4.13,[20] we show how CVN 77 start date and build period affect CVN 77 costs

[20]As in the preceding graphs of this type, this and the following graphs assume purchase of CFE no earlier than the shipyard start date.

when the other work in the yard includes a CVX start in 2006. (Recall from Chapter Three that starting CVX later than 2006 will not sustain a 12-ship fleet.) We assume for the purpose of this analysis that a CVX would generate labor demand similar to that of the Nimitz class. As we would expect, there is little difference from the curves in Figure 4.10 for CVN 77 start dates and schedules that result in most of the work being conducted before CVX is started.

It is interesting to compare the costs in Figures 4.10 and 4.13, because the difference between them is a measure of CVX's influence on the cost effects we attribute to CVN 77. Figure 4.10 shows gains and losses relative to the baseline (6.5-year build starting in 2002) caused by varying CVN 77 start date and build period, without CVX; Figure 4.13 shows analogous gains and losses with CVX. In Figure 4.14, we show the remainder when a cost gain or loss graphed in Figure 4.10 is subtracted from the analogous cost gain or loss in Figure 4.13. A positive difference indicates that the effect of introducing CVX would be to decrease the gain (or increase the loss) associated with varying the start date and build period of CVN 77, relative to the base case. For example, when CVX is omitted, total shipyard costs are $320 million less if a 7.5-year CVN 77 build is started in 2000 than if the baseline case holds, i.e., if a 6.5-year build starts in 2002 (Figure 4.10). If CVX is included, the analogous cost differential is $270

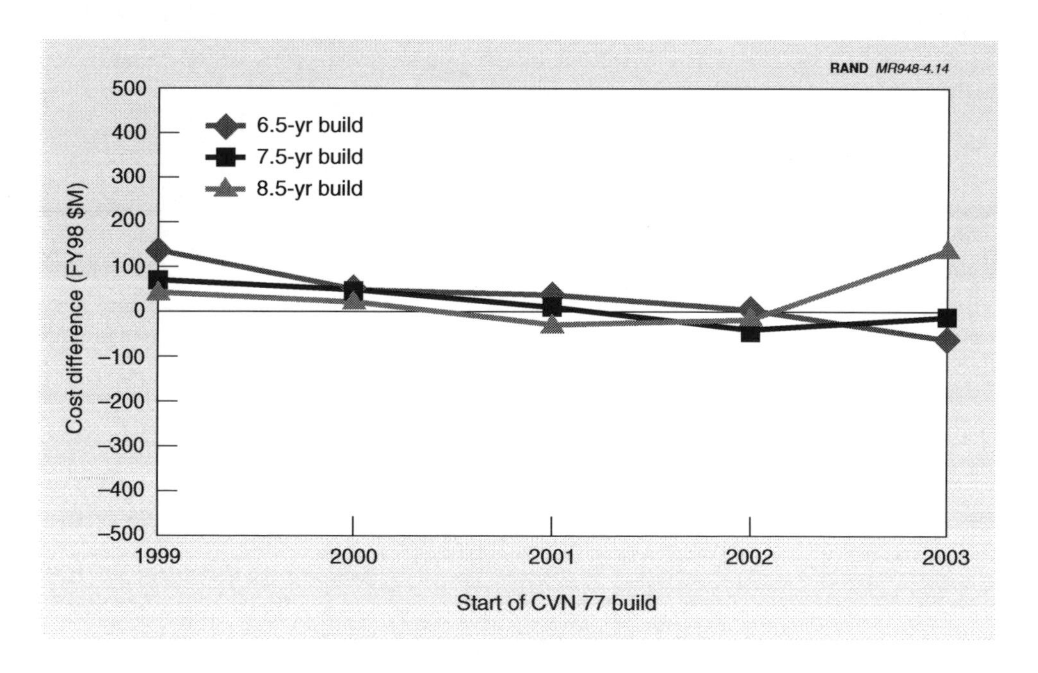

Figure 4.14—Effect of CVX on Extra Cost or Savings Associated with Varying CVN 77 Start Date and Build Period

million (Figure 4.13). Thus, introducing CVX makes the 7.5-year build beginning in 2000 look $50 million worse, and that $50 million is graphed on the 7.5-year line at 2000 in Figure 4.14.

Figure 4.14 illustrates that, regardless of the start date or build period, the results given in Figure 4.10 are not much affected when CVX is taken into account. For earlier starts, the effect of CVX is to increase the relative cost (i.e., decrease the savings), but it increases it less so as starts occur later. Following the lines to the right, we see that, at some point, the longer build periods begin to overlap the build period of CVX too much and the cost relative to the base case increases again.

The point about small CVX effects is particularly true if CVN 77 is started earlier but the delivery date remains at 2008. Here again, we are interested in the diamond at 2002, the square at 2001, and the triangle at 2000. The effects of CVX on these three CVN 77 schedules are essentially zero—they are certainly less than our estimating error.

CONCLUSIONS AND A CAVEAT

We have shown that, for any given CVN 77 build period, total shipyard costs fall as the start date is moved earlier than 2003 (which is about as late as CVN 77 can be started if a 12-ship fleet is to be sustained). Also, for any given start date between 1999 and 2003, costs generally fall with increasing build period. For a fixed delivery date of 2008 (that currently planned), as start date is moved earlier, costs fall—by some $400 million if CVN 77 is begun in 2000. These conclusions hold whether CVX is built at Newport News Shipbuilding (beginning in 2006) or not.

We have arrived at our cost estimates by moving CVN 77 labor and costs around against a fixed background of other demands. However, these other demands may not materialize as projected and additional sources of work may not materialize. Furthermore, the shipyard and the Navy, in a search for ways to minimize costs across all projects, may move around some of the demands already anticipated. During the course of this research, for example, the distribution of submarine work changed.

What if one of the anticipated tasks changes—the refueling/complex overhaul for CVN 70, beginning in September 2005, for example? Figure 4.15 shows what happens to the results in Figure 4.7 when that overhaul is started a year earlier or a year later than planned. The effects are small if CVN 77 is started by 2002, but they are quite large for later starts. For example, according to Figure 4.7, starting a 6.5-year CVN 77 build in 2003 will cost $310 million more than start-

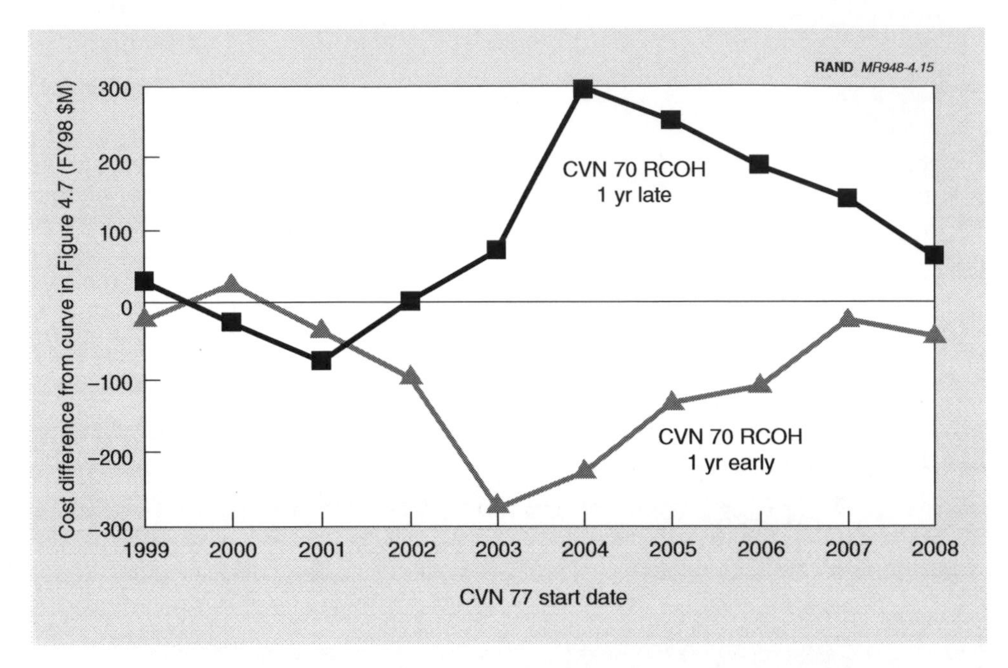

Figure 4.15—Response of Figure 4.7 Results to Changes in the Schedule of Another Project (CVN 70 RCOH)

ing it in 2002. But, according to Figure 4.15, starting the CVN 70 overhaul a year earlier than is assumed in Figure 4.7 would reduce the effect shown in Figure 4.7 by $280 million. Thus, there would be little change, and the major cost penalty that we report above for delaying CVN 77's start one year would be almost obliterated by a 1-year change in the schedule of just one of the other projects in the yard.

This analysis confirms what many public- and private-shipyard and government officials told us about the effects of relatively modest schedule changes on total shipyard costs. It also demonstrates the importance of coordinating and understanding the effects of new-production and maintenance schedules on the shipyard.

Changes in project scheduling at Newport News could also influence, in unpredictable ways, schedules and costs for Navy projects in other shipyards. For example, expending funds to get work on CVN 77 started earlier may mean postponing funding of work in another yard, which may result in costly swings in the labor-force demand at that yard. Estimating such effects would require an analytic tool that could be applied to more than one yard.

Chapter Five

HOW CVN 77 CONSTRUCTION AFFECTS NUCLEAR-COMPONENT VENDORS

Since 1988, contractors in the naval nuclear-propulsion industrial base have been realigning their workforces and facilities to match the workload reduction associated with nuclear-powered submarine and surface ship production programs and force structures. During the early 1990s, when the Seawolf submarine program terminated after three submarines,[1] uncertainty about the timing of future orders and anticipated low production rates raised concerns that the remaining producers of naval nuclear components might close down or cease making those components. In 1993, the award of CVN 76 and the initiation of the NSSN program helped to stabilize the industry, albeit at a much smaller scale. Currently, the nuclear industrial base is sized to support submarine and carrier refuelings and production of at least one new attack submarine per year.

In this chapter, we examine the implications for nuclear vendors of producing CVN 77, the last of the Nimitz-class ships. We show that it matters little to the survival of the nuclear industrial base whether CVN 77 is produced or not—a result that obviates the need to assess the effects of various start dates. However, the industrial base for certain carrier nuclear components faces a long production gap, which raises challenging issues that need to be settled if CVX is nuclear. Although that gap is not strictly related to decisions regarding CVN 77 construction, we take the opportunity to discuss this important potential problem here.

This analysis draws on our discussions with and data provided by the nuclear industry, the Naval Nuclear Propulsion Directorate, the Naval Sea Systems Command, and officials of the Office of the Secretary of Defense (OSD). We begin the chapter with an overview of the naval nuclear industrial base, then go on to issues for CVN 77 and CVX.

---

[1]The Seawolf class was projected to comprise as many as 12 submarines (*Selected Acquisition Report*, December 1988).

NAVAL NUCLEAR-PROPULSION INDUSTRIAL BASE

Research, development, and manufacture in support of the nuclear Navy are carried out by major corporations under contract to the government and by subcontractors that supply hardware support and technical expertise to the government and the prime contractors. In this section, we discuss the importance and prospects of suppliers who manufacture critical nuclear components, including the

- reactor cores
- control-rod drive mechanisms
- pumps, pipes, and fittings
- instrumentation and control equipment
- valves and auxiliary equipment
- heavy reactor-plant components (reactor vessels, closure heads, core barrels, steam generators, pressurizers).

Table 5.1 lists the manufacturers of these components.

It is essential for the Navy's nuclear ship programs that the naval nuclear industrial capability survive. Nuclear-system manufacture requires high standards for component manufacturing and quality assurance, specialized facilities for fabrication and testing, and a highly qualified and skilled workforce. Nuclear reactor core manufacture for the Navy is even more specialized and demanding. Naval nuclear reactors are smaller than commercial reactors, use highly enriched fuel, and must operate for decades without being replaced or having major maintenance. They experience frequent power variations, are required to meet quietness and shock criteria, and are designed to operate in proximity to humans. The means of meeting this kind of demand cannot be replaced quickly or cheaply, if it is practical to replace it at all.

The prospects of the naval nuclear industrial base are less than robust, as is starkly illustrated by the dwindling number of critical-component suppliers to the Navy (Table 5.2). *All the key suppliers for reactor-plant components are sole-source.* With no domestic civilian orders and declining naval orders, the nuclear field is no longer commercially appealing.[2] The large capital investment

[2]For a discussion of the commercial nuclear industry during its early years, see Robert Perry et al., *Development and Commercialization of the Light Water Reactor, 1946–1976*, Santa Monica, Calif.: RAND, R-2180-NSF, June 1977.

Table 5.1

Key Nuclear Suppliers

Nuclear Component	Supplier	Location
Nuclear cores	BWX Technologies Naval Nuclear Fuel Div.	Lynchburg, VA
Heavy components	BWX Technologies Nuclear Equipment Division	Barberton, OH
Control-rod drive mechanisms	Marine Mechanical Corporation	Cleveland, OH
Pumps, pipe, and fittings	Westinghouse Electro-Mechanical Division	Cheswick, PA
	BW/IP International, Byron Jackson Pump Division	Long Beach, CA
	Taylor Forge	Paola, KS
Instrumentation and control equipment	SPD Technologies	Philadelphia, PA
	Eaton Pressure Sensors Division	Bethel, CT
	Lockheed Martin Tactical Defense Systems (LMTDS)	Archibald, PA
	Lockheed Martin Information Systems	Orlando, FL
	Peerless Instrument Corporation	Elmhurst, NY
	Imaging & Sensing Technology Corp.	Elmira, NY
	Northrop Grumman Power/Control Systems Division	Baltimore, MD
	Eaton Cutler-Hammer	Milwaukee, WI
	Power Paragon	Anaheim, CA
Valves and auxiliary equipment	Target Rock Corp.	East Farmingdale, NY
	Hamill Manufacturing	Trafford, PA

SOURCE: Supplement to the Naval Propulsion Directorate, Naval Sea Systems Command, *March 3 Report on Preservation of the U.S. Nuclear Submarine Capability,* Naval Nuclear Industrial Base report, November 10, 1992; updated through interview with Naval Nuclear Propulsion Directorate.

needed and the low probability of achieving a satisfactory return on investment will probably discourage any new firms from entering the nuclear market.

LAST-OF-CLASS IMPLICATIONS FOR THE INDUSTRIAL BASE

In 1969, Congress authorized the Navy to procure one complete shipset of reactor-plant heavy-equipment components as a backup to minimize the risk of either delaying construction of nuclear carriers or laying up those carriers,

Table 5.2

History of Naval Nuclear-Component Suppliers

Component	Component Suppliers 1960s	1970s	1980s	1990s
Reactor Cores	B&W UNC CE M&C West.	B&W UNC	B&W UNC	BWXT
Heavy Equipment	B&W A-C AOS FW CE West. Alco	B&W AC CW CE Aero SW	B&W PCC[a] CE SW	BWXT
Control Rod Drive Mechanisms	TRW VARD M-S	TRW R/LSI[a]	TRW BFM[a]	MMC[a]
Main Coolant Pumps	West. GE	West.	West.	West.

NOTE:

AC	Allis Chalmers	GE	General Electric
Aero	Aerojet	M&C	Metals and Controls
AOS	A.O. Smith	M-S	Marvel-Schelber
B&W	Babcock and Wilcox renamed BWX Technologies (BWXT)	PCC	Precision Components Corp.
BFM	Barry, Frank, & Murray	UNC	United Nuclear Corp.
BWXT	See B&W	R/LSI	Royal/Lear Siegler Inc.
CE	Combustion Engineering	SW	Struthers Wells
CW	Curtiss-Wright	West.	Westinghouse
FW	Foster Wheeler		

[a]Successor in same facility to company listed on same line to the left.

once in service.[3] Because of the long lead times needed for manufacture of these large nuclear-propulsion-plant components, the production backup set has typically been used to construct the next CVN. The components manufactured with the advance procurement (AP) funds for that ship have then replaced the production backup.

This practice will be followed for CVN 77 to the extent that the spare shipset will be used in constructing that ship. However, only a partial replacement spare shipset of reactor-plant components is planned for procurement with CVN 77

[3]Department of Defense Appropriations Bill, U.S. House of Representatives, H.R. 18707, 1969.

funds. For FY00, the Navy's budget includes AP funds to procure reactor cores; control-rod drive mechanisms; pumps, pipes, and fittings; instrumentation and control equipment; and valves and auxiliary equipment. Funds are *not* included to replace the spare long-lead heavy equipment—such as steam generators, reactor vessels, core barrels, closure heads, pressurizers, and some supporting equipment—which leaves the spare shipset incomplete.[4] Therefore, there will be no CVN 77 work for the heavy-equipment vendor, regardless of whether or when CVN 77 is built. This vendor will complete its current CVN work in FY00 and is downsizing to reduce costs as it focuses on meeting the equipment demand for smaller nuclear-submarine components.

As for nuclear components other than the heavy equipment, the cores and the instrumentation and control equipment will be replaced at some point in the carrier's life, either during reactor refueling or in some other maintenance action (see Table 5.3 for the core procurement schedule). Pumps, pipes, and fittings, as well as valves and auxiliary equipment, are needed for ongoing submarine production. These requirements will sustain the industrial base for the components other than heavy equipment.

In short, CVN 77 does not affect the heavy-equipment sector of the nuclear industrial base. However, if CVX is nuclear, there are reasons to be concerned on

Table 5.3

Core Procurement Schedule for New Construction and Refueling

Ship	Fiscal Year																		
Type	98	99	00	01	02	03	04	05	06	07	08	09	10	11	12	13	14	15	16
SSBN[a]	1		1		1		1		1		1		1						
CVN[b]		1	2[c]			1	1	1	1		1		1	1	1		1		1
NSSN[d]	1	0	1	1	1	1	2	2	2	2	2	2	2	2	2	2	2	2	2

[a]Assumes a force level of 14 Tridents.

[b]CVN estimates assume ships have lifetime operating tempo similar to that of USS *Nimitz*.

[c]CVN 77 new-construction core assumes Advance Procurement funds in FY00.

[d]Based on one new attack submarine per year, plus refuelings.

---

[4]Using the backup heavy equipment for construction of CVN 77 without initiating replacement will leave the Navy with no backup to support constructing CVN 77 and operating the fleet. Damage to any of the major components would delay ship construction for years while a replacement component is completed. Cost increases due to construction delays and rushing construction of replacement components could be hundreds of millions of dollars. Similarly, a need to replace this equipment in an operating carrier could result in a ship "immobilized a considerable time until replacement components can be obtained" (Admiral Rickover, congressional testimony before the Committee on Armed Services, United States Senate, Ninetieth Congress, Second Session, March 1968).

behalf of the heavy-equipment manufacturer. These issues are addressed in the following section.

MANUFACTURING HEAVY EQUIPMENT FOR CVX[5]

BWX Technology's Nuclear Equipment Division (NED) is the sole source for heavy-equipment components. NED is currently using less than 30 percent of its facilities, which occupy 107 acres and have a fabrication area of 1.7 million square feet and office area of 230,000 square feet. As the naval reactors work has drawn down, NED has also shed workers. Figure 5.1 shows the substantial drawdown in employment levels from 1988 to the present, along with planned future employment levels. NED's forecast of 322 employees by 2001 is based on production of one NSSN per year.

Carrier components account for a substantial part of the workload at NED. The work for 1 Nimitz-class CVN is equivalent to that of 7 submarines; the CVX, if nuclear, is expected to be equivalent to 4–6 submarines (Table 5.4). Current employment levels may increase, depending on the size of a CVX nuclear-reactor plant and its resulting work requirement.

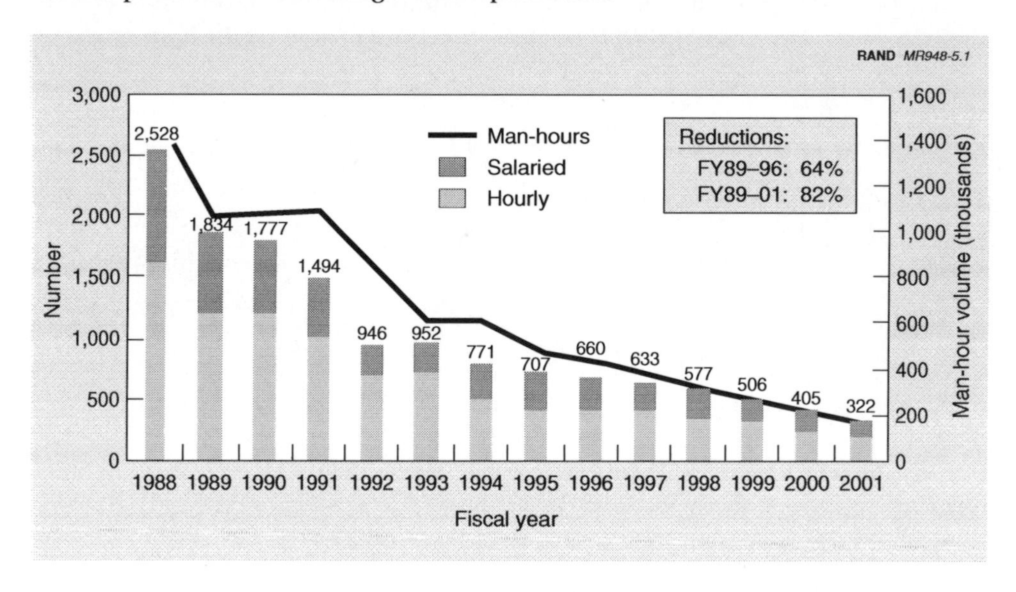

Figure 5.1—Shrinkage of Workforce to Match Volume Reduction, BWX Technology, Nuclear Equipment Division

---

[5]This section draws on a December 6, 1996, meeting at BWX Technology's Nuclear Equipment Division (NED), as well as on many pieces of correspondence and phone discussions among RAND staff, Naval Nuclear Propulsion Directorate staff, and NED personnel.

Table 5.4

CVN and NSSN Heavy-Equipment Components, for Comparison

Characteristic	CVN Nimitz Class	NSSN
Fabrication man-hours	1.1M	0.15M
Number of components	16 large	6 small
Approximate weight	2,000 tons	200 tons

The production of both submarine and carrier heavy-equipment components may be hampered by the policies used to downsize the workforce at NED. Union rules have forced NED to make termination decisions based on low seniority, resulting in a workforce in which few employees are younger than 45 years old,[6] as Figure 5.2 shows. The current workforce at NED is highly experienced, to the point where there is a virtual absence of junior employees, as shown in Figure 5.3. And, as NED continues to draw down its workforce, the employees at the left edge of the distribution in Figure 5.3 (i.e., the younger employees) will be the ones terminated.

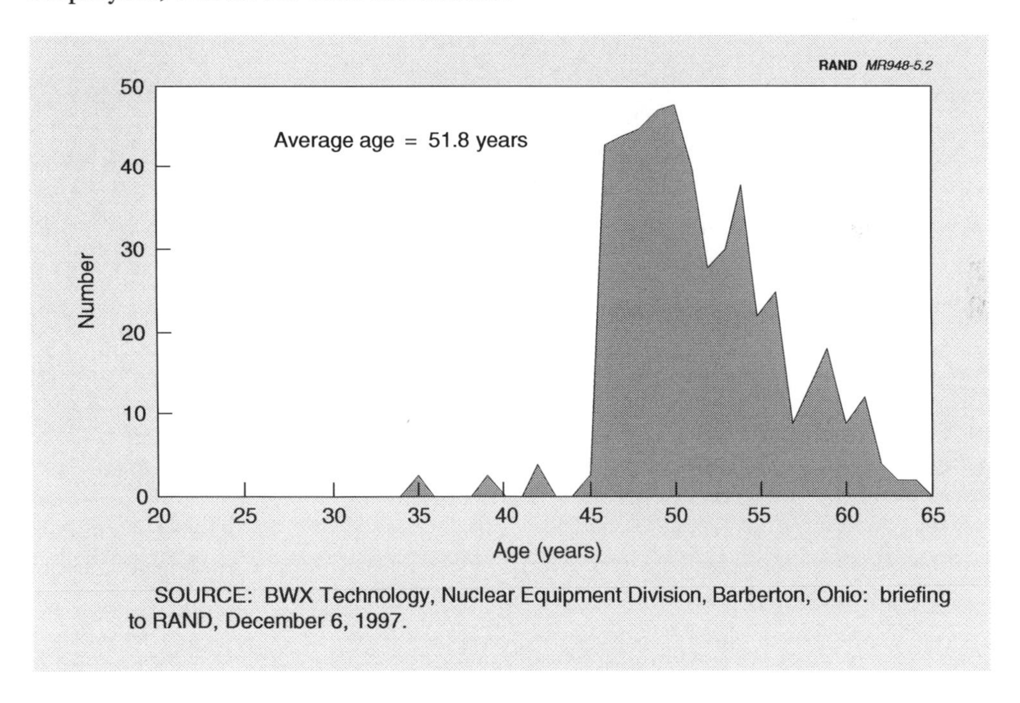

Figure 5.2—Age of Hourly Workforce, BWX Technology, Nuclear Equipment Division, as of November 14, 1996

[6]A few employees in their thirties or early forties have the technical skills required for operating modern computer-aided design and manufacturing (CAD/CAM) equipment.

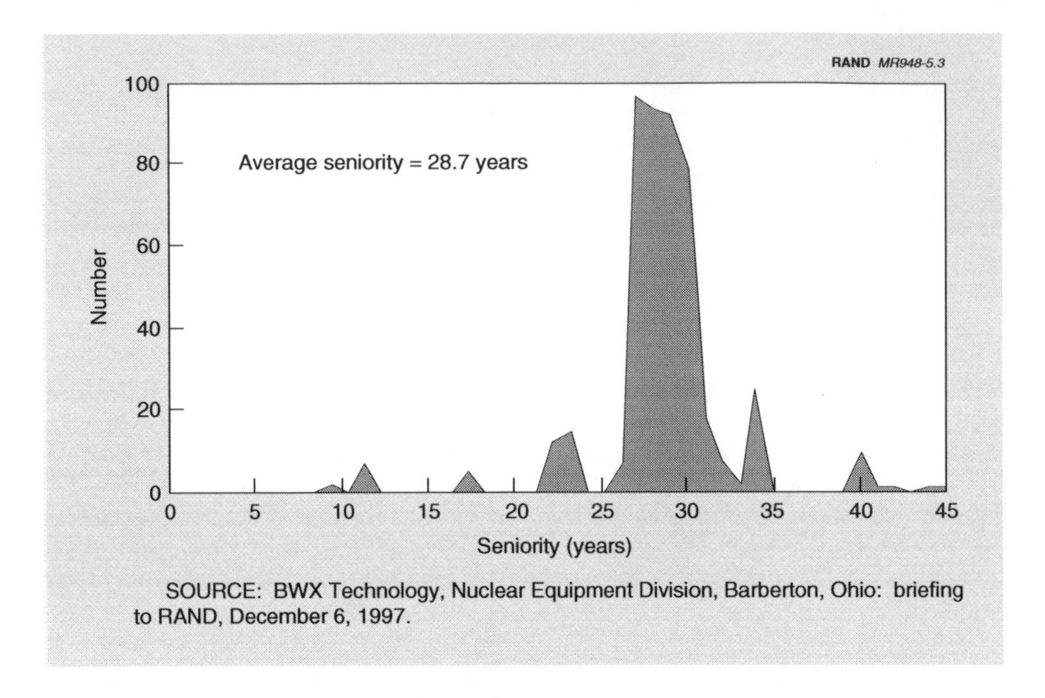

SOURCE: BWX Technology, Nuclear Equipment Division, Barberton, Ohio: briefing to RAND, December 6, 1997.

Figure 5.3—Seniority of Hourly Workforce, BWX Technology, Nuclear Equipment Division, as of November 14, 1996

The naval nuclear community is aware of this situation and is working with NED to facilitate its transition to the next generation of workers. The major issues are how many new personnel should be hired and trained, and when that hiring and training should occur.

At NED, the CVX nuclear-versus-conventional-propulsion decision extends beyond workforce considerations to equipment and facility decisions. The current carrier work will be completed by FY00. Even now, some fabrication and welding equipment is no longer needed for the remaining carrier work. NED must now make decisions on the disposition of that equipment and also of facilities dedicated to carrier-component construction. NED must either mothball equipment and facilities and commit to some continuing overhead costs or sell them off.

The Navy's planned schedule for new heavy-equipment design, development, and construction assumes that a nuclear-propulsion decision is made in FY00 and ship authorization in FY06 (see Figure 5.4).[7] NED recommends the

[7]Traditionally, advance-procurement funding for long-lead nuclear components is provided two years prior to ship authorization. However, because of the long lead time needed for heavy-

schedule shown in Figure 5.5, which provides for start of heavy-equipment fabrication in FY00. The firm estimates its recommended schedule will cut $40 to $60 million from the workforce and facility retrenching and the subsequent rebuilding that would occur over the 4-year hiatus assumed in the notional schedule.[8] NED states that making the propulsion decision on CVX by FY99 will allow it to make more-efficient decisions on equipment, facilities, and personnel and thus to realize the $40–$60-million savings.

The point about rushing the schedule is worth elaborating. For the Nimitz class, six to seven years has been the average manufacturing time for components such as reactor vessels, pressurizers, steam generators, closure heads, and core barrels. That period is longer than the interval between contract award and shipyard need. Thus, for CVX, fabrication of heavy-equipment components would need to begin in advance—well in advance—of hull construction, even if all production processes are up and running.

Suppose that CVX 78 has the same shipyard construction schedule as CVN 76 and that its heavy nuclear-system equipment takes the same time to manufacture as that of the Nimitz class. Fabrication of some components would then have to start five to seven years in advance of the shipyard's work (see Figure 5.6), or 14 years in advance of the ship's delivery. Granted, CVX 78's nuclear equipment is expected to require less work to build than CVN 76's. However, CVN 76 is the ninth ship of a class, and some inefficiencies must be allowed for first-of-a-kind components, including uncertainties in lead time for both design and fabrication.[9] In addition, the CVN 76 schedule takes advantage of an experienced, continuously employed workforce; the use of new hires for nuclear CVX components might slow the schedule even more.

CONCLUSIONS

The production start date and schedule for CVN 77 have little effect on nuclear vendors. In fact, it matters little to the viability of the nuclear vendors whether CVN 77 is produced at all. Manufacturers of light equipment and reactor cores will survive for the foreseeable future with the work associated with NSSN construction and with submarine and carrier refuelings. And, because an already-

---

equipment manufacture, Naval Reactors has told the Navy and OSD that advance-procurement funding for heavy equipment must be provided no later than FY02 if CVX is nuclear-powered.

[8]Letter from BWX Technology to John Birkler at RAND, dated January 14, 1997.

[9]For the first ship of a class, time must also be allowed for developing and testing the components.

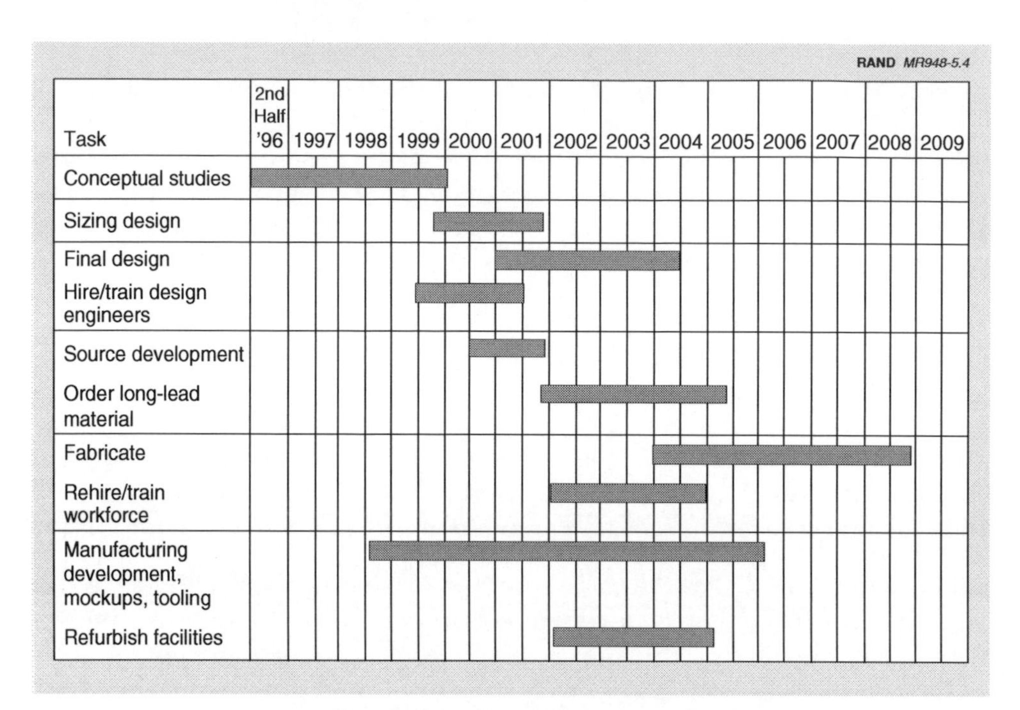

Figure 5.4—Schedule for CVX (Nuclear) Heavy-Equipment Components (FY06 CVX 78 Shipyard Start)

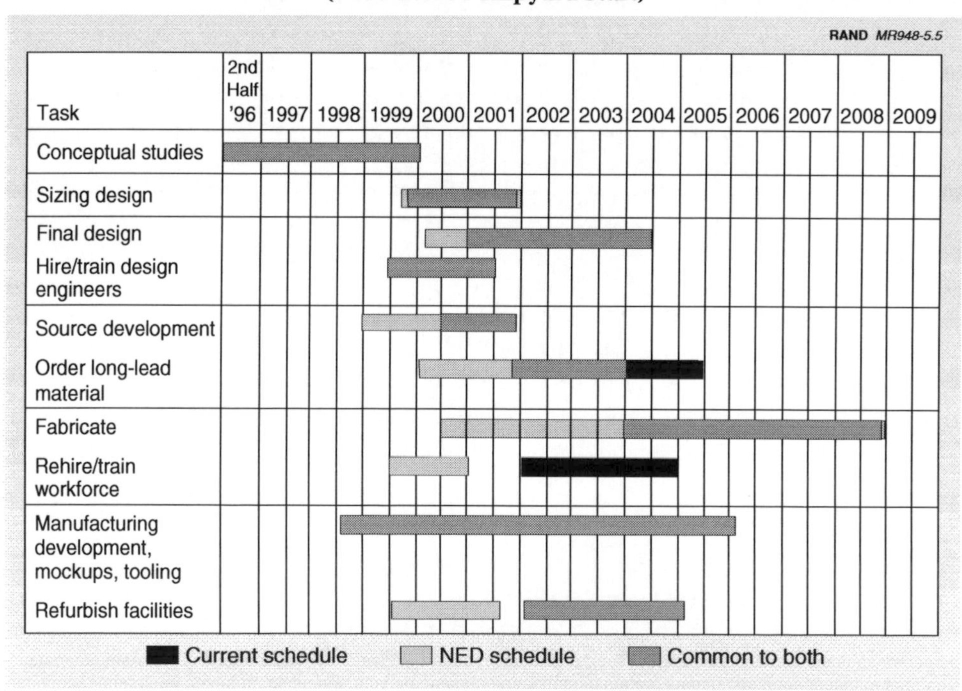

Figure 5.5—BWX Technology's Recommended Schedule for Heavy-Equipment Components and Current Schedule (FY06 CVX 78 Shipyard Start)

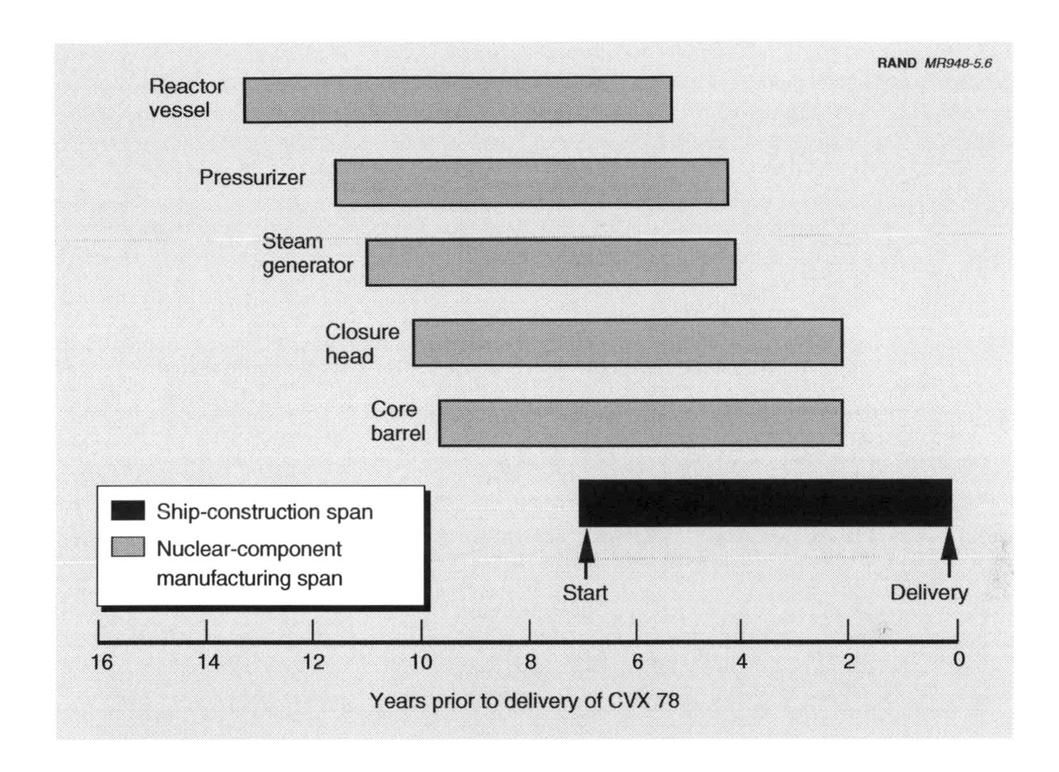

Figure 5.6—Shipyard Need Dates and Manufacturing Spans for Heavy-Equipment Components, Based on CVN 76 and Nimitz-Class Experience

constructed spare set of heavy equipment will be used on CVN 77, there will be no CVN 77 work for the sole remaining manufacturer of heavy equipment. This manufacturer—the Nuclear Equipment Division of BWX Technology—must survive only on NSSN construction.

Although CVN 77 is not critical to the survival of naval nuclear vendors, the timing of the propulsion decision for CVX does have major implications for NED, because the remaining carrier work will come to an end in FY00. If the Navy adheres to its current plan of beginning heavy-equipment fabrication for CVX in 2004 (if CVX is nuclear), NED will mothball or sell construction equipment and replace workers only as needed for NSSN work. If a decision to make CVX nuclear-powered is made by FY00, fabrication could begin earlier, which would avoid the costs of retrenching and reconstituting its production capability. NED estimates this would save $40 to $60 million.

Chapter Six

ISSUES RELATED TO SUPPLIERS OF NONNUCLEAR COMPONENTS

In addition to the shipyard itself, the industrial base for aircraft carrier construction includes a large complex of firms providing the thousands of individual carrier components that the shipyard does not fabricate itself. These components range from such standard items as fasteners, to somewhat specialized items such as pipes and valves, to such highly specialized items as catapults and arresting gear. They include those bought by and supplied directly to the shipyard (contractor-furnished equipment [CFE]) and those bought and supplied to the shipyard by the government (government-furnished equipment [GFE]). In this chapter, we focus on CFE suppliers.[1]

We sought answers to three questions about the industrial base that would be called on to supply nonnuclear components for CVN 77 and subsequent carriers:

- Will qualified suppliers be available and willing to support construction of the next aircraft carrier?
- When will commitments for long-lead items be needed in order to meet the postulated delivery date of 2008 for CVN 77?
- What are the cost consequences of various dates when orders are placed with major CFE vendors for the next carrier (CVN 77)?

Some broad characteristics of the vendor base impinge on our approach to answering these questions. First, over 2,000 firms are supplying CFE for CVN 76, which is now under construction; these firms account for about $800 million worth of products. It is obviously not practicable to examine each of these firms

[1]A large portion of the GFE consists of electronic equipment (radars, communications equipment, etc.), which, while adapted in design for carriers, is based on a broad set of products having other military and commercial uses. Furthermore, GFE that is modified or updated to meet evolving mission needs is, at times, backfitted to some or all operating carriers; CFE, on the other hand, is seldom replaced. Thus, typical suppliers of GFE have a much broader and more continuous market for their products than do typical suppliers of CFE.

in the detail devoted to the shipyard itself and to the firm supplying the heavy nuclear equipment. Fortunately, half the total CFE business (in dollars) is provided by about two dozen firms, and 40 percent is provided by ten firms (see Figure 6.1).

Second, an overarching problem for most of the suppliers, especially those providing items that are unique to carriers, is the intermittence of carrier work. The usual interval between carrier starts is four or five years, and there will apparently be six years between the starts of CVN 76 and CVN 77. It takes some suppliers two years or less to satisfy the demand from a carrier, so they must have other lines of work. In fact, the several suppliers of high-cost items produce a variety of goods and services; the products made for aircraft carriers generally represent a narrow, specialized niche in their product base. Since each firm's product base (and its management approach) evolves, so does its ability to deliver items for carrier construction. Because of the evolving diversity of CFE suppliers, our research questions can be meaningfully answered only for the aggregate of firms; furthermore, the answers may change as time passes.

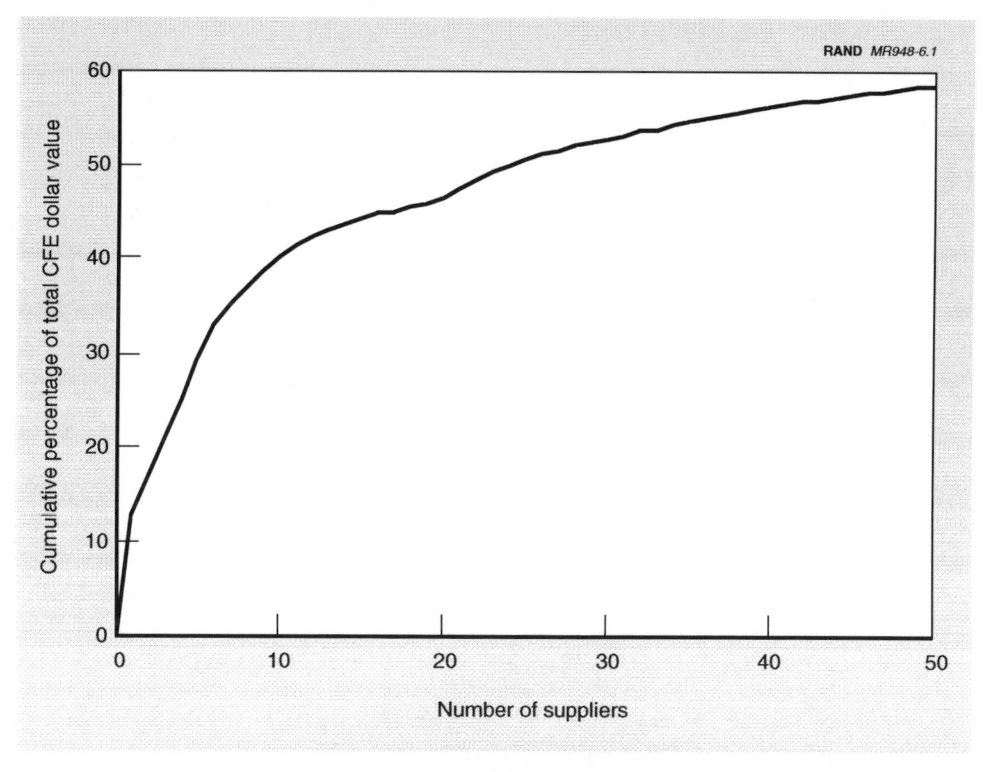

Figure 6.1—Total Dollar Value of Contractor-Furnished Equipment in Relation to Number of Suppliers

In this analysis, we have made no effort to examine any firm in sufficient detail to permit postulating a specific course of action to ensure an efficient supply of that firm's products. Instead, we have tried to understand the range of factors and situations affecting the majority of firms supplying products for aircraft carrier production. In answering the questions posed above, we offer an aggregate assessment of that industrial base and some broad strategies the Navy and the shipyard might follow to minimize problems in obtaining the necessary products in the future.

AVAILABILITY OF QUALIFIED SUPPLIERS

CFE must often be designed especially for carrier use; therefore, there is an advantage to obtaining equipment from an experienced supplier. But, as we pointed out above, many firms supplying CFE cannot survive on carrier orders alone, because carrier construction is intermittent. There is thus always the risk that a qualified supplier will come to regard its carrier work as expendable—particularly as the defense industry continues to downsize and consolidate. Firms are acquired by new owners who may regard carrier-specific product lines as less important than the old ones did. These new owners may then be less inclined to allocate the necessary production resources to satisfying an order for carrier equipment.

To understand better the degree to which a continuous vendor base may be threatened, we discussed the matter with officials of Newport News Shipbuilding and with several firms making critical products for the current class of aircraft carriers. Although hardly an exhaustive survey, its results were so uniform as to persuasively support a conclusion: Barring some unforeseen set of circumstances, *we do not anticipate major problems in obtaining commitments from present suppliers to furnish their products for the next aircraft carrier (CVN 77),* assuming it is started within the 2000–2002 time frame. If the next carrier is delayed further, our confidence in the availability of present, proven suppliers begins to decrease, if only because of uncertainties regarding the overall industrial environment and the situations of specific firms that far in the future.

SCHEDULING PROCUREMENT OF LONG-LEAD ITEMS

Many aspects of our industrial-base analysis are driven by timing and scheduling factors. One of the most important of these is the timing of orders for long-lead items, items that require a long time for the supplier to produce and particularly those that must be installed early in the construction of a carrier. Orders for such items are usually placed one or two years before work begins in earnest on carrier construction. Here, we examine the timing of such orders and how that timing affects the overall carrier-construction schedule.

It is useful to review the time intervals required for construction of recent carriers. In Figure 6.2, we show the overall build period for the Nimitz-class carriers, divided into four sequential intervals:

1. Contract award date to start of fabrication (SF) of hardware in the shipyard.
2. Fabrication start to keel laying. (Beginning with CVN 71, the construction process changed to one in which major modules are assembled in an area beside the dry dock, then placed in the dock and assembled into the ship proper. For CVN 71 and later, "keel laying" is equivalent to placing the first pre-assembled module into the dry dock.)
3. Keel to launch, where *launch* denotes moving the ship from the dry dock to a nearby pier for further fitting out.
4. Launch to delivery of the completed carrier, ready for commissioning and shakedown operations.

Since we are concerned with the order, delivery, and integration of major components requiring long lead times and that are assembled into modules, we want to examine the overall ship-construction time from keel laying to delivery of the ship. This interval for the Nimitz-class carriers to date is shown in Figure 6.3. For the last six ships, the construction method has been the same and

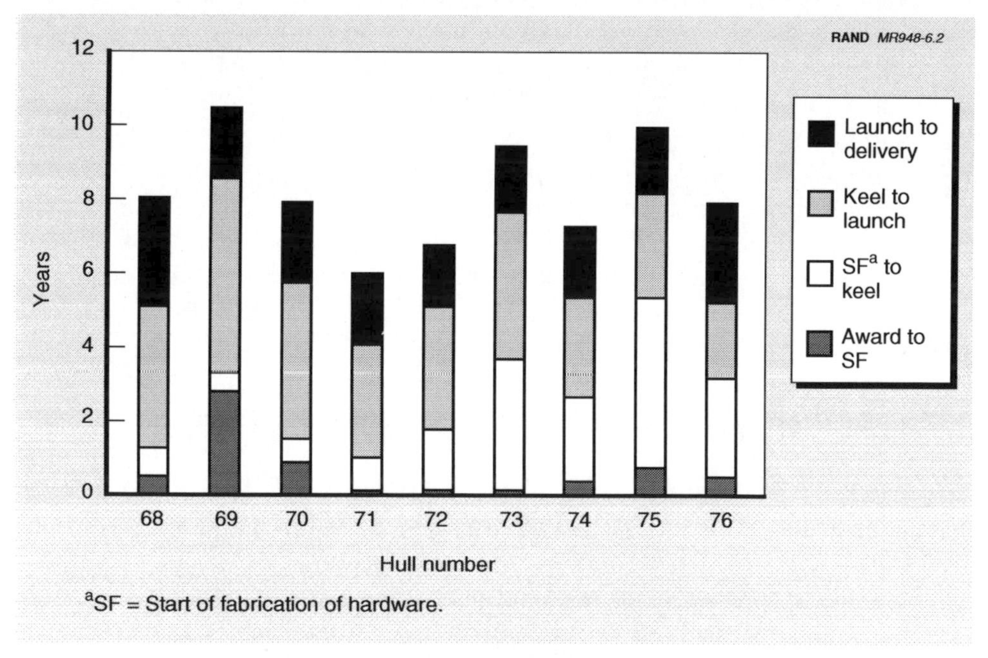

Figure 6.2—Recent Ship-Construction Times (Award to Delivery)

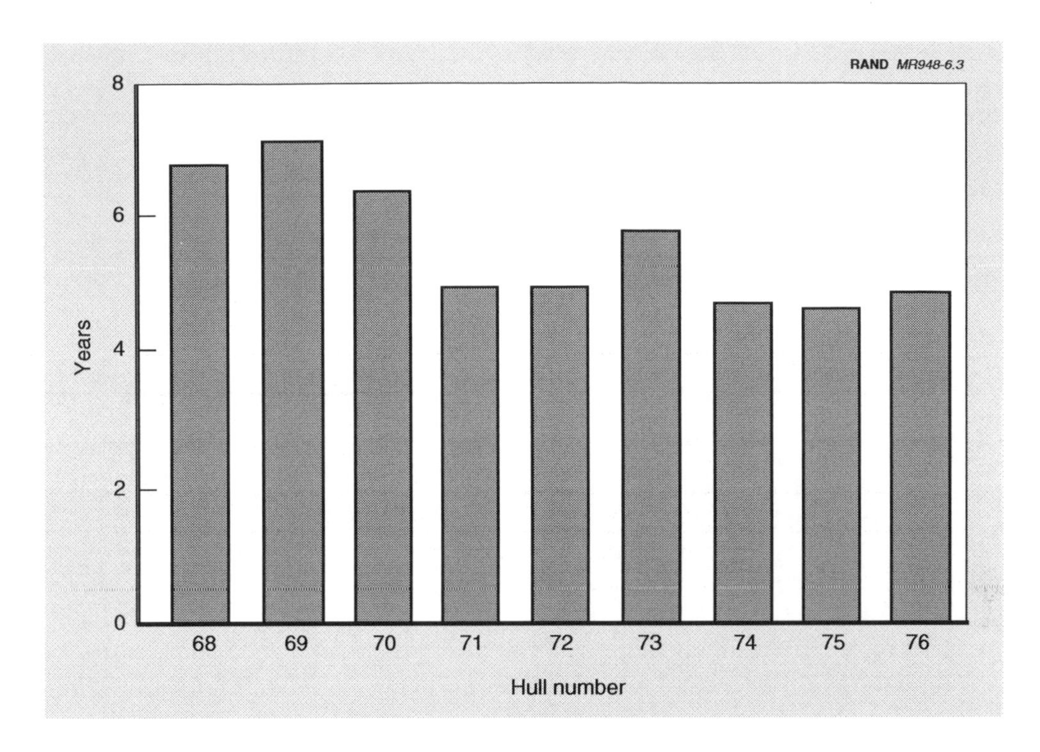

Figure 6.3—Recent Ship-Construction Times (Keel to Delivery)

the time from first module (keel) to final delivery has been remarkably constant: about five years, plus or minus a few months. While that interval could be compressed somewhat, shipyard officials believe that about five years from keel to delivery (equivalent to about eight years from contract award) is close to optimal, and that compression would increase costs and risk. We therefore adopt a period of five years as the nominal duration for the keel-to-delivery interval.

Major items such as the main propulsion-drive turbines and speed-reduction gears, which require 24 to 30 months to assemble, must be installed relatively early in the shipbuilding process. In general, 24 to 30 months is a good estimate for items that require special manufacturing processes and that are ordered only at extended intervals, so that the supplier cannot sustain a continuous manufacturing activity.[2] If such items are to be delivered to the shipyard at or near the start of the 5-year keel-to-delivery interval, then orders for those major items must be placed seven to eight years before scheduled ship delivery. Such

[2]Times for specific items in any given plant may vary outside this range, depending on the timing of other work in the vendor's plant.

timing corresponds well to the overall ship-construction times (contract award to delivery) shown in Figure 6.2 and is therefore an important element in establishing the practical delivery date for an aircraft carrier. Thus, if the Navy wants CVN 77 delivered in 2008, orders for some of the major ship components should be placed by 2000.

COST CONSEQUENCES OF VENDOR ORDER DATES

Most of the vendors supplying high-cost components to CVN 76 will complete their work in 1997 or 1998. Any gap between that time and component orders for CVN 77 will cause some loss in the readiness of the vendor to fill the next order. Certain factors lead to a "restart cost," which must be added to the price of the next shipset of products from each vendor. Among those factors are normal turnover in staff, which leads to a loss of workers having recent experience in manufacturing processes; tooling, which gets shunted aside to make room for other work; and second- and third-tier vendors, which suffer some attrition. In general, the restart-cost penalty tends to increase most rapidly in the first year or two after completion of the prior order; then the rate of increase diminishes as the restart cost approaches the cost required after an extended gap in production.

An understanding of the restart-cost penalty for vendor-supplied components and materials should be of help in planning the overall schedule and in budgeting funds for the next aircraft carrier. Some limited but useful information on this topic can be gleaned from estimates provided by a few of the major vendors during discussions with Newport News Shipbuilding during the summer and fall of 1996.

Vendors representing about 45 percent of the total value of components and supplies for CVN 76 production provided estimates of the price for the next shipset of products, according to when the order was placed, with order times ranging from 1998 to 2002.

The aggregate results are shown by the solid line in Figure 6.4, which we consider a lower-bound estimate of total vendor restart costs because it does not include many vendors producing the remaining 55 percent of total CFE value.

Conducting a similarly detailed survey of the many vendors in that remaining 55 percent was impractical. To get some estimate of the likely restart costs for those other vendors, we first observed that about one-fourth of the large vendors surveyed indicated no change in price as a function of order date (within the range of dates examined). It seems plausible that a somewhat larger fraction of the remaining vendors might be supplying standard products and that their price quotes would not be very sensitive to order date. Therefore, we as-

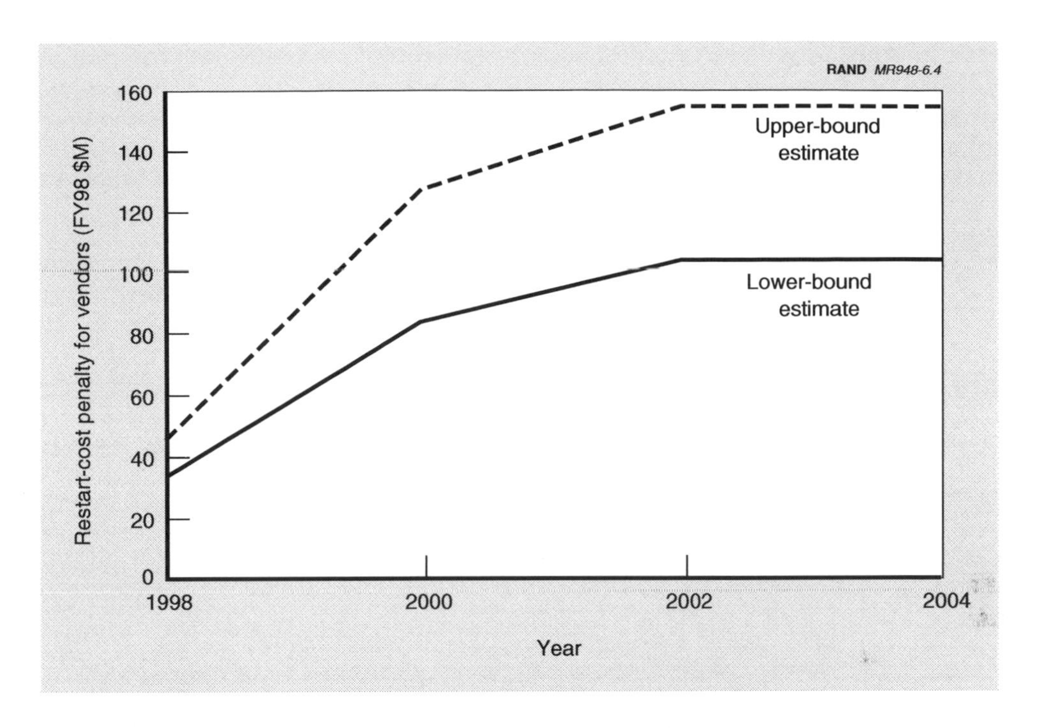

Figure 6.4—Vendor Cost Increment Attributable to Production Gap

sumed that one-half of the remaining vendors would experience restart costs proportional to those of the high-cost products, leading to the estimates shown by the dashed line in Figure 6.4.[3] We believe the dashed line represents a rough upper bound of the range of such estimates for all vendor-supplied products needed for CVN 77 construction. We further assumed that both the lower- and upper-bound estimates will be essentially invariant for orders placed after 2002. While any given vendor's situation can change quickly, we believe the aggregate results indicate with some reliability the cost changes to be expected with later placement of orders for major CVN 77 components and material.

CONCLUSIONS

We answer the three questions we posed at the beginning of this chapter as follows:

---

[3]The increment represented by the dashed line is one-half the value of the points on the solid line. We realize this assumes a 50/50 missing/surveyed split instead of 55/45. It further assumes that, for the half of the missing vendors with proportional costs, those costs are proportional to the costs from all the surveyed vendors (instead of, e.g., the three-quarters with nonzero changes). However, the imprecision is consistent with the very rough nature of the one-half estimate.

- Qualified suppliers should be willing and able to support construction of the next aircraft carrier if construction is started within the 2000–2002 time frame.
- To meet the postulated CVN 77 delivery date of 2008, the Navy will need to commit funding to some long-lead items by 2000.
- Compared with costs expected for a CVN 77 start in 2002, CFE costs can be reduced by $20 to $30 million if ordered by 2000 and by $70 to $110 million if ordered by 1998. (Here, the lower ends of the ranges are derived from the lower-bound estimates shown in Figure 6.4 and the upper ends are derived from the upper-bound estimates.)

Chapter Seven

RESEARCH AND DEVELOPMENT TO SAVE COSTS ON FUTURE CARRIERS

We now turn to our final set of research questions regarding CVN 77 design and construction:

- First, are there technologies and applications-engineering processes not now used by builders of Navy ships that might reduce the life-cycle costs of CVN 77, earlier Nimitz-class ships, or the CVX class?
- If there are, and in view of the potential for cost reduction, how much of an R&D investment should be made to permit the adaptation of those technologies and processes for CVN 77?

Nimitz-class aircraft carriers are very expensive ships to build, own, and operate. The procurement cost for CVN 77 will be more than $5 billion. Each Nimitz-class carrier has an average annual operations and maintenance cost of approximately $240 million, slightly less than half of which is for shipyard availabilities. A little less than a third of the annual cost is for the pay and allowances of the approximately 3,000 enlisted and officer ship's company personnel assigned to each carrier. Even a small percentage reduction in annual costs can have a major impact on the annual ship construction, Navy (SCN), military personnel, Navy (MPN), and operations and maintenance, Navy (O&MN) budgets.

CVN 77 will be the tenth, and last, of the Nimitz-class nuclear aircraft carriers. Although the basic design of the class is over 30 years old, no two Nimitz-class carriers are exactly the same. Each carrier incorporates changes from its previous sister ship. At times, the changes—the redesigned island and bulbous bow added to CVN 76, for example—are fairly significant; most changes, however, are minor, incorporating the latest equipment or weapon systems. The vast majority of the changes incorporated over the past 30 years have been in operational areas, to satisfy new mission requirements or to increase the survivability of the ship against new enemy threats. Few, if any, of the design changes address reductions in maintenance or other areas of life-cycle cost or attempt to improve the quality of life for the Navy personnel on board the ship.

CVN 77 offers an opportunity to identify and incorporate more-modern production processes and technologies to reduce life-cycle costs, improve operational availability, and enhance shipboard quality of life. Also, CVN 77 can serve as a transition ship: Any changes made to it can provide a foundation, or test bed, for the future CVX class of aircraft carriers. Changes that reduce the number and cost of shipboard personnel or of shipyard availabilities might also be backfitted to previous Nimitz-class ships. However, design changes to CVN 77 require R&D funding to identify any improvements and to incorporate those improvements in the basic design plans and construction processes for the ship.

To identify innovative production processes, modern technologies, and new application-engineering approaches, we reviewed the available literature and interviewed several firms that build commercial ships. Commercial shipbuilders are motivated by the competitive forces of the marketplace to reduce costs and increase the availability of their products to their eventual owners. Modern cargo ships operate with crews of less than 20, and modern cruise ships are built to be operational 50 weeks a year. We cannot, of course, expect aircraft carriers to meet such goals, but we anticipated that something might be learned from the commercial sector about reducing crew and maintenance needs.

Ideally, such lessons would be learned from commercial U.S. shipbuilders. Unfortunately, almost no commercial shipbuilding base survives in the United States. (Indeed, if it had, the lessons we sought might already have been learned by and applied within the military sector.) For this reason, we interviewed European shipbuilders, particularly those building cruise ships, which, like aircraft carriers, must sustain large numbers of persons on board.[1]

In this chapter, we describe the results of our literature reviews and interviews with foreign firms, then present and demonstrate an analytic approach for determining appropriate levels of R&D funding for CVN 77.

---

[1]Appendices F, G, H, and I present the results of our interviews with various foreign shipbuilders and with the naval defense organizations of Great Britain and France. The foreign firms include the Kvaerner shipyard in Glasgow, Scotland; Chantier d'Atlantique in Paris, France; the Kvaerner Masa-Yard in Helsinki, Finland; and the Fincantieri shipyard in Monfalcone, Italy.

IDENTIFYING ADAPTABLE PRODUCTION PROCESSES AND TECHNOLOGIES

The interviews with European commercial cruise-ship builders identified no single new technology, no "silver bullet" they use to reduce various elements of a ship's life-cycle cost. Rather, commercial firms apply many modern production processes and existing technologies, each aimed at controlling some element of production or operating cost. Although the effect of a process may be minor when considered separately, the cumulative effect of the set of technologies and processes is a significant cost reduction.

Production Processes

European shipbuilders concentrate their in-yard efforts primarily on steel fabrication and the construction of the basic hull form, relying on turnkey subcontractors to install many of the ship's "hotel" functions (for example, berths, food service, laundry, and waste management). Prefabricated passenger cabins (complete with plumbing and bed linens), modular anchor-handling machinery, the laundry, the galley, gambling rooms, and ship bridges are all examples of turnkey systems now being installed in commercial cruise ships. Typical of this approach is the service provided by the Finnish companies Lopart Systems and Electrolux, which have joined to create a complete design-build team for food-service systems on many cruise ships. Working with the shipbuilder, Kvaerner Masa, at the beginning of the ship's design, Lopart and Electrolux identify the space and service requirements for the food services, then design and build the systems at their plants and deliver and install them, virtually complete, in the ship.

Another example is the construction of modular cabins by a Kvaerner subsidiary. Built in Piikkiö, Finland, approximately 100 miles west of the Helsinki shipyard, the cabins are transported to the shipyard on semitrailers. Kvaerner Masa then slides each cabin into the appropriate area on the ship, tacks it down, and connects the electrical, plumbing, and air-conditioning systems—all in approximately 10 hours.

The commercial shipbuilders use an open architecture for the hotel functions on board the ship, specifying only form, fit, and function. They entertain bids from various subcontractors, because they see an advantage to using companies that build, for example, many cabins or galleys each year versus trying to accomplish that function less frequently with their own employees. In their opinion, using experienced subcontractors not only lowers costs (they typically quoted 20 to 30 percent in cost reductions), but also reduces the risks to the shipbuilder and provides better overall quality.

To make outsourcing successful, ships are designed for modular construction and for incorporation of a wide range of commercial equipment. Most of that equipment is obtained from suppliers having a broad business base, which helps ensure that selected equipment has wide usage and, therefore, wide availability. Helpful not only during ship construction, the widespread availability of much of the equipment at most seaports throughout the world also helps during ship operations.

Commercial shipbuilders also use various processes to reduce shipyard overhead costs. Many are moving toward just-in-time delivery of materials to reduce the cost of financing, the need for large laydown and materials-storage areas, and interim maintenance of the materials and parts. The Kvaerner Masa yard, which uses approximately 25,000 tons of steel per year for the hull and structural parts, has arranged with its steel producer to provide the plates already cut to size, bent to shape, blasted, coated, and with ends prepared for welding. Because of the just-in-time supply philosophy, no more than 100 steel plates are in the yard at any one time, awaiting assembly into modules.

In relationships with customers, the commercial shipbuilders use fairly flexible management practices and contract-change provisions. Procedures for handling changes when they do occur are simple and straightforward. Overall, the relationships can be characterized as very cooperative and long-term. For example, the Kvaerner Masa shipyard is nearing completion of a series of eight ships for Carnival Cruise Lines.

Technology

Commercial shipbuilders use numerous existing technologies to reduce costs. For example, several manufacturers, such as Courtauld Coatings (International Paint Company) are now offering paints formulated to dry more quickly and require fewer coats. The new paints are claimed to have superior abrasion resistance and to last longer than conventional paints, properties that reduce the frequency, and cost, of repainting.[2] These paints have been used by shipyards in Europe, Japan, and Korea on over 300 ships, including commercial tankers and other complex vessels.

Other manufacturers, such as Jotun, offer competing systems with similar advantages. Jotun has improved anti-bottom-fouling paints, fast-drying shop primers geared to the just-in-time delivery of steel, and improved systems for use in such harsh environments as holding tanks. On the horizon are highly

---

[2]A major cost during the availabilities of Nimitz-class ships is the painting of tanks and voids. For example, *Nimitz* (CVN 68) had costs of almost $90 million (FY98 $) for repainting tanks and voids during the first 20 years of its life.

advanced "surface-modification" systems, which, unlike paints that depend on adhesion to stay in place, are chemically bonded to surfaces and are therefore permanent. Passive anti-fouling systems are being developed to create a hydrophilic surface that prevents fouling by appearing (to the fouling organisms) as no surface at all. Other surface-modification systems incorporate enhanced water-shedding properties or provide a permanent barrier to oxygen and moisture.

Technology innovations are available in other areas as well. Firms such as Deerberg Systems (Germany), Norsk Hydro (Norway), and FOX Pollution Packers (United Kingdom) provide commercial shipboard waste-management systems—shredders, compactors, pulpers, water extractors, sewage processors, waste-oil combustors, and incinerators—that offer the range of approaches from complete onboard containment and destruction of waste material to the pulping, shredding, and discharge approach adopted recently by the U.S. Navy. All systems are constructed in modules so that they can easily be interconnected or operated independently of each other.

A Danish innovation by LR Industries cuts and shapes piping to specification at the factory. The piping is insulated with mineral wool and plastic foam for service at temperatures ranging from –200 to +450 degrees centigrade. The insulation, which is constructed of multiple layers of very durable and watertight materials, provides a strong outer surface that is very resistant to mechanical damage. Pipe hangers are fixed to the outer layer of insulation to prevent heat leaks and hanger-maintenance problems. Moisture detectors are also available that reveal the existence and location of pipe leaks inside the insulation. This piping-insulation system is currently used in steam and condensate generators on chemical and product tankers, fuel and heated oil lines on bulk carriers, and cargo vapor lines on liquid-petroleum-gas (LPG) tankers.

Required for making oil tanks inert, membrane generators that can deliver up to 1500 cubic meters per hours of nitrogen (with up to 5 percent oxygen)—believed to be the largest production rate for membrane nitrogen generators achieved to date—are available from Permea Maritime Protection, a Norwegian company.

Composite materials are also slowly becoming more common in commercial ship construction. Composites Engineering of Great Britain is supplying fire-resistant composite vinyl-ester glass-reinforced plastic deck grating, dosed with carbon (to improve electrical conductivity and prevent buildup of static electricity), to the builder of seven chemical tankers. Flexible shaft couplings of carbon-fiber composite are being manufactured by Centa Antriebe, a German company. The composite shafts are lighter and can span greater distances without bearings than can their steel counterparts.

Finally, various commercial components are being used to reduce crew requirements on board commercial cargo and tanker ships. These include remote sensor systems, closed-circuit cameras, electromechanical valves, and other electronic subsystems.

Applied Engineering

Further cost-savings or operations-enhancing advances in commercial shipbuilding can be loosely classified as "applied engineering." For example, a consortium of German firms, led by Germanischer Lloyd, is in the midst of a 5-year project to develop the tools needed for improving life-cycle structural design of ships. Particular areas of attention are vibration prediction and modeling, loading effects, fatigue strength, collapse behavior under extreme loads, fabrication effects on structural performance, and monitoring of structures during service. The American Bureau of Shipping (ABS) has a similar project for designing and evaluating hull structures, which is called SafeHull.

Along these lines, the Japanese ship-classification society, Nippon Kaiji Kyokai (ClassNK) has developed a computer-based ship-assessment program, PrimeShip, which provides design, construction, operation, and maintenance guidelines to maximize the service life of ships. These guidelines cover essentially every aspect of ships, including design, propulsion, hydrodynamics and maneuverability, and scrapping.

A consortium of organizations from Korea, Denmark, Finland, and Norway has developed a Windows NT–based diagnostic system for the engine room of ships to analyze vibration and particle counts, along with such conventional inputs as temperature and pressure, permitting prediction of failure in time to correct problems. The first system is being installed on a new Korean container ship.

Applied engineering has also focused on ship-propulsion systems. Many European and Asian ship designers are working to improve the propulsion efficiencies and vibration performance of propellers and associated skegs and rudders. For example, SI-Shipping AB of Sweden designed a new twin-shaft chemical tanker with asymmetric skegs that impart a rotational field in the wake against the propeller rotation. The propellers are highly skewed and lightly loaded. Overall, the system delivers a superb propeller efficiency of nearly 0.8. The bulky skegs also provide added space for cargo. Kappel, a Danish firm, offers fin-tip propellers, which are based on work by the U.S. National Aeronautics and Space Administration (NASA) that led to fin tips on aircraft wings. These propellers claim an improvement of 3 to 5 percent over the efficiency of ordinary propellers.

Finally, modern cruise ships must operate at a variety of speeds and have good fuel economy across the speed range. Hybrid diesel/diesel-electric and turbine/diesel-electric propulsion plants are providing the needed flexibility. The new P&O Cruise Line ship *Oriana*, constructed by the German builder Meyer Werft, is a twin-shaft ship, 70,000 tons gross, with two main propulsion diesels (one 6-cylinder and one 9-cylinder) clutched and geared to each shaft. Each propulsion diesel also drives an attached electric alternator that can serve as a motor to boost shaft horsepower during high-speed transits. Normal electric power is provided by four diesel-generator sets. All together, 11 possible combinations of diesel and electric motors can drive each shaft. The entire propulsion and auxiliary plant of the ship is controlled by a Siemens automated control system. Comparable levels of propulsion flexibility and automation are incorporated in other new cruise ships from the European industry.

In summary, in our literature search and interviews with the European commercial shipbuilders, we identified a wide range of techniques that are now being used to reduce various elements of life-cycle cost or to improve operational performance. Further research is needed to understand the magnitude of the potential cost savings, the applications to naval ships, and the rate of return (or time to recoup the initial investment) appropriate to the various production processes, new technologies, or applied-engineering techniques.

We next discuss an analytic method for determining appropriate levels of R&D funding and provide initial estimates of such funding for CVN 77.

ESTIMATING R&D INVESTMENT TO REDUCE LIFE-CYCLE COSTS

As indicated above, many opportunities for reducing production, maintenance, and personnel costs present themselves. Collecting data on and analyzing the payoff expected from any of these opportunities are beyond the scope of our study. However, it is relatively easy to demonstrate that there is potentially a very significant aggregate payoff and that it is probably worth spending several hundred million dollars to pursue some of the modern commercial practices identified above.

In this section, we present such a demonstration for two major categories of maintenance and operations costs: scheduled depot maintenance activities and enlisted-personnel pay (ship's company only; no air wing personnel) for Nimitz-class carriers.

Costs of Scheduled Availabilities and Enlisted Crew

The Navy VAMOSC data system for ships provides the most comprehensive view of ship operating and support (O&S) costs.[3] The system consists of 130 elements and subelements, organized into four major categories. Table 7.1 shows annual costs per ship for each of these categories, and, for the two largest, for several subcategories. Costs are based primarily on time series (i.e., breakdown by year) for Nimitz-class carriers.[4] For reasons to be explained below, experience-based estimates for two large subcategories—scheduled overhauls and fleet modernization—had to be modified on the basis of other information. Those two subcategories, together with the enlisted-crew portion of the personnel subcategory, account for 78 percent of O&S costs. In the following paragraphs, we expand on our derivation of the costs for these categories.

Table 7.1

Nimitz-Class Operating and Support Cost Breakdown

Category	Annual Cost per Ship (FY98 $M)	Percentage of Total
Direct Unit Costs	**105**	**43**
Personnel	86	35
Officers	11	4
Enlisted	75	31
Materials	13	5
Purchased Services	6	3
Direct Intermediate Maintenance	**1**	**0**
Direct Depot Maintenance	**128**	**53**
Scheduled Overhauls	78	32
Non-Scheduled Overhauls	7	3
Fleet Modernization	37	15
Other	6	3
Indirect O&S	**9**	**4**
Total	**243**	**100**

[3]VAMOSC stands for Visibility and Management of Operating and Support Costs. All military services initiated VAMOSC data systems in the mid- to late-1970s. The Navy maintains two major VAMOSC systems—one for aircraft and one for ships.

[4]Appendix J shows O&S cost time series for individual ships.

Table 7.2 presents the schedule and estimated costs for the new Nimitz-class Incremental Maintenance Program (IMP).[5] As shown, following a 6-month shakedown cruise and a 4-month postshakedown availability (PSA), the carrier's life consists of a set of 18-month cruise periods separated by planned incremental availabilities (PIAs). Every third PIA involves placing the ship in dry dock; the others are accomplished along a pier. The docking PIAs (DPIAs) are planned to take 10.5 months each; the others are planned to take 6 months. At midlife, the carrier goes through a refueling/complex overhaul.[6]

Table 7.2

Nominal Nimitz-Class Availability Schedule and Estimated Costs

Event	Duration (mo)	Cumulative Time (mo)	Cumulative Time (yr)	Date for CVN 77 (month-year)[a]	Estimated Cost (FY98 $M)[b]
Commissioned				Jul-08	
Cruise	6.0	6.0	0.5	Dec-08	
PSA	4.0	10.0	0.8	May-09	50
Cruise	18.0	28.0	2.3	Oct-10	
PIA 1A	6.0	34.0	2.8	May-11	120
Cruise	18.0	52.0	4.3	Oct-12	
PIA 1B	6.0	58.0	4.8	May-13	120
Cruise	18.0	76.0	6.3	Oct-14	
DPIA 1	10.5	86.5	7.2	Sep-15	200
Cruise	18.0	104.5	8.7	Mar-17	
PIA 2A	6.0	110.5	9.2	Sep-17	135
Cruise	18.0	128.5	10.7	Mar-19	
PIA 2B	6.0	134.5	11.2	Sep-19	135
Cruise	18.0	152.5	12.7	Mar-21	
DPIA 2	10.5	163.0	13.6	Jan-22	235
Cruise	18.0	181.0	15.1	Aug-23	
PIA 3A	6.0	187.0	15.6	Jan-24	150
Cruise	18.0	205.0	17.1	Jul-25	
PIA 3B	6.0	211.0	17.6	Jan-26	150
Cruise	18.0	229.0	19.1	Aug-27	
DPIA 3	10.5	239.5	20.0	Jun-28	265
Cruise	18.0	257.5	21.5	Dec-29	
PIA 4A	6.0	263.5	22.0	Jun-30	150
Cruise	18.0	281.5	23.5	Dec-31	
RCOH	32.0	313.5	26.1	Aug-34	2000

[5]The Incremental Maintenance Program replaces the Engineered Operating Cycle for Nimitz-class aircraft carriers (*Incremental Maintenance Program Manual*, January 1, 1997). The nominal IMP schedule is also shown in the *Aircraft Carrier Continuous Maintenance Program (ACCMP) Manual*, a document that is the master plan for both conventionally and nuclear-powered aircraft carriers. A draft version of this document was provided to RAND to support the present study. A final version may be available by the time this report is published.

[6]CVN 68, *Nimitz*, arrived at Newport News in May 1998 for its midlife refueling and complex overhaul, the first of its class to undergo the process. Based on CVN 68 fuel use, the RCOH for Nimitz-class ships will occur at a ship age of approximately 23 years. This timing—and thus the timing of ship retirement—may vary with the OPTEMPOs of individual ships.

Table 7.2—continued

Event	Duration (mo)	Cumulative Time (mo)	Cumulative Time (yr)	Date for CVN 77 (month-year)[a]	Estimated Cost (FY98 $M)[b]
Cruise	6.0	319.5	26.6	Feb-35	
PSA	4.0	323.5	27.0	Jun-35	50
Cruise	18.0	341.5	28.5	Dec-36	
PIA 2A	6.0	347.5	29.0	Jun-37	135
Cruise	18.0	365.5	30.5	Dec-38	
PIA 2B	6.0	371.5	31.0	Jun-39	135
Cruise	18.0	389.5	32.5	Dec-40	
DPIA 2	10.5	400.0	33.3	Oct-41	235
Cruise	18.0	418.0	34.8	May-43	
PIA 3A	6.0	424.0	35.3	Oct-43	150
Cruise	18.0	442.0	36.8	May-45	
PIA 3B	6.0	448.0	37.3	Oct-45	150
Cruise	18.0	466.0	38.8	May-47	
DPIA 3	10.5	476.5	39.7	Mar-48	265
Cruise	18.0	494.5	41.2	Sep-49	
PIA 4A	6.0	500.5	41.7	Mar-50	150
Cruise	18.0	518.5	43.2	Sep-51	
PIA 4B	6.0	524.5	43.7	Mar-52	150
Cruise	18.0	542.5	45.2	Sep-53	
DPIA 4	10.5	553.0	46.1	Jul-54	265
Cruise	18.0	571.0	47.6	Jan-56	
PIA 5A	6.0	577.0	48.1	Jul-56	150
Cruise	18.0	595.0	49.6	Jan-58	

NOTE: The PSAs are not shown in the IMP; when the first PSA is included, there is not enough time for the IMP's PIA 4B preceding the midlife RCOH, so it is omitted here. Numbering of PIAs after the RCOH begins with 2 because the level of effort—and cost—of the first PIA after the RCOH is expected to resemble PIA 2 in the first half of the ship's life.

[a]Beginning in 2008.

[b]We estimated these costs by multiplying man-day values from the IMP by cost per man-day (resulting from the ratio of costs expressed in FY98 dollars to manpower for completed CVN repair and modernization).

At the time of this study, Navy planning for the transition to the IMP covered the schedule and the estimated manpower for the new set of availabilities. However, cost estimates had not yet been developed. To satisfy the requirements of this study, RAND developed the cost estimates shown in the last column of Table 7.2 by determining the ratio of costs (expressed in FY98 dollars) to manpower for completed CVN repair and modernization work during availabilities.[7] To arrive at the estimated costs in the last column of Table 7.2, we

[7]Historical cost and manpower data were provided by PERA-CV, the Navy's aircraft-carrier planning and engineering organization, located near Puget Sound Naval Shipyard in Bremerton, Washington.

multiplied the resulting values for cost per man-day by the man-day values from the IMP.[8]

To calculate enlisted-crew costs and savings, we again turn to VAMOSC, which shows that such costs for the Nimitz class are at least $75 million per year. We say "at least" because VAMOSC does not cover all categories of cost,[9] and because for each Navy person assigned to a ship, it is estimated that approximately two additional Navy personnel are required ashore to handle all the support operations. Therefore, in using $75 million per year in the following analysis, we are being conservative.

Estimated Investment Considering CVN 77 Savings Alone

A net present value (NPV) analysis is frequently used to evaluate the financial attractiveness of an investment. In general terms, an investment involves spending an amount of money in anticipation of a future payoff. Both the initial outlay and the future returns may be spread over several periods of time. NPV analysis calculates the value of this stream of outlays and returns from the point at which the initial outlay is made, working from the assumption that the present value of a dollar becomes less the farther in the future it is spent or saved.[10] Usually, future dollars are discounted at a constant rate per year. The NPV is the sum of the discounted values of the outlays and returns.

With regard to incorporating cost-savings improvements in CVN 77 (or the rest of the Nimitz class), we assume that expenditures occur in FY98 and that returns in the form of reduced O&S costs begin when the ship starts operations, FY09.[11] Figure 7.1 is a plot of the availability costs from Table 7.2; it assigns

---

[8]PERA-CV reviewed these results and agreed that they constitute reasonable estimates at this time.

[9]The costs included in the Enlisted Manpower element of VAMOSC are cost of services of active-duty Navy enlisted personnel assigned to the ship, as reported by Defense Finance and Accounting Services–Cleveland Center from the Joint Uniform Military Pay System (JUMPS). "This includes base pay, allowances, other entitlement and government contributions to FICA [Federal Insurance Contributions Act] and SGLI [Servicemen's Group Life Insurance]. This element does not include the indirect cost of trainees, unassigned personnel, permanent change of station, prisoners, patients, enlisted subsistence, etc." U.S. Naval Center for Cost Analysis, *Navy Visibility and Management of Operating and Support Cost (Navy VAMOSC): Data Reference Manual for Individual Ships Report,* Arlington, Va., April 30, 1997, p. II-3.

[10]This value does not include the effects of inflation. All costs in this report are in constant FY98 dollars—i.e., inflation is ignored—which is appropriate, because future inflated costs will be paid for with future inflated dollars. However, even ignoring inflation, most people would rather have a dollar now than a dollar a year from now, if for no other reason than a dollar invested now should be worth more than that dollar a year from now.

[11]The current planning is for CVN 77 to be commissioned in July 2008. It will thus have three months of operations during the end of FY08, a period that covers a portion of the shakedown cruise. The first depot maintenance activity is the PSA, which occurs at the end of the shakedown cruise and is in FY09.

each cost to the year during which funding would be required for that availability. The total of the costs shown is $5.545 billion.

The question we are addressing here is how much investment should be considered to achieve a reduction in the cost stream shown in Figure 7.1 and in that represented by the $75 million per year in enlisted-crew costs given in Table 7.1. The general answer is that the United States should not invest more than the discounted present value of that reduction or savings. For a more specific answer, one needs to know how much can be saved and what discount rate to use.

We begin in Figure 7.2 by showing the NPV of the *total* scheduled availability costs and that of the enlisted-crew costs as valued in FY98 dollars, for a range of discount rates. Using the current interest rate recommended by the Office of Management and Budget for discounting constant-year dollars, which is 3.6 percent,[12] we get an NPV of the total availability cost of $1.7 billion.

How much of this $1.7 billion is a candidate for savings? Discussions with a wide range of naval personnel indicate that studies of savings opportunities are

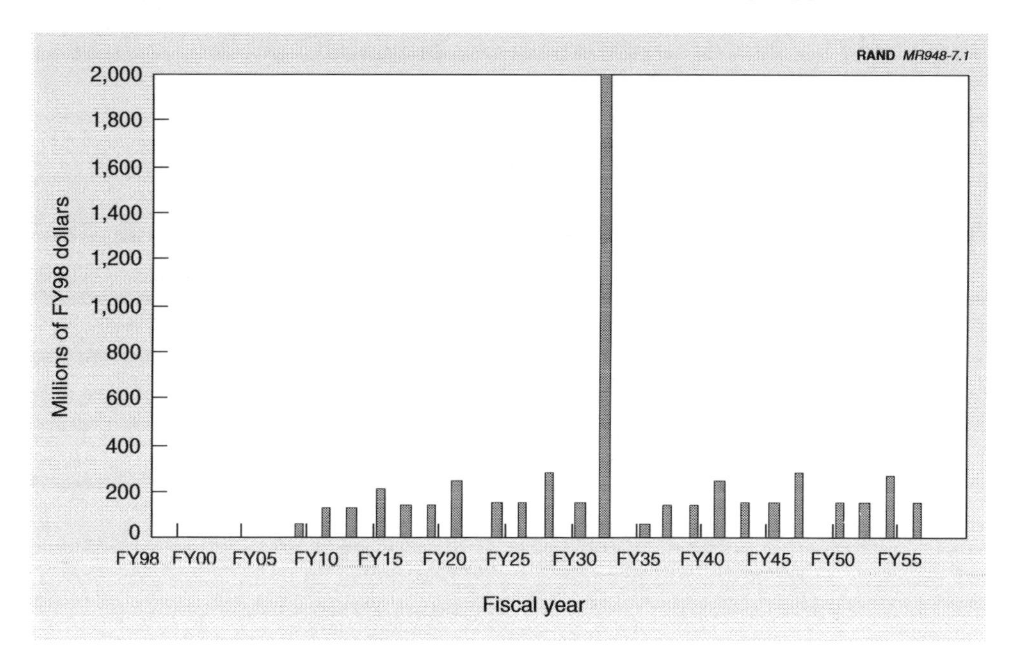

Figure 7.1—Anticipated Costs of Scheduled CVN 77 Availabilities, by Year

---

[12]Office of Management and Budget, *Memorandum for Heads of Executive Departments and Establishments (Subject: Guidelines and Discount Rates for Benefit-Cost Analysis of Federal Programs)*, Circular A-94 (revised), October 23, 1992, Appendix C (revised February 1997).

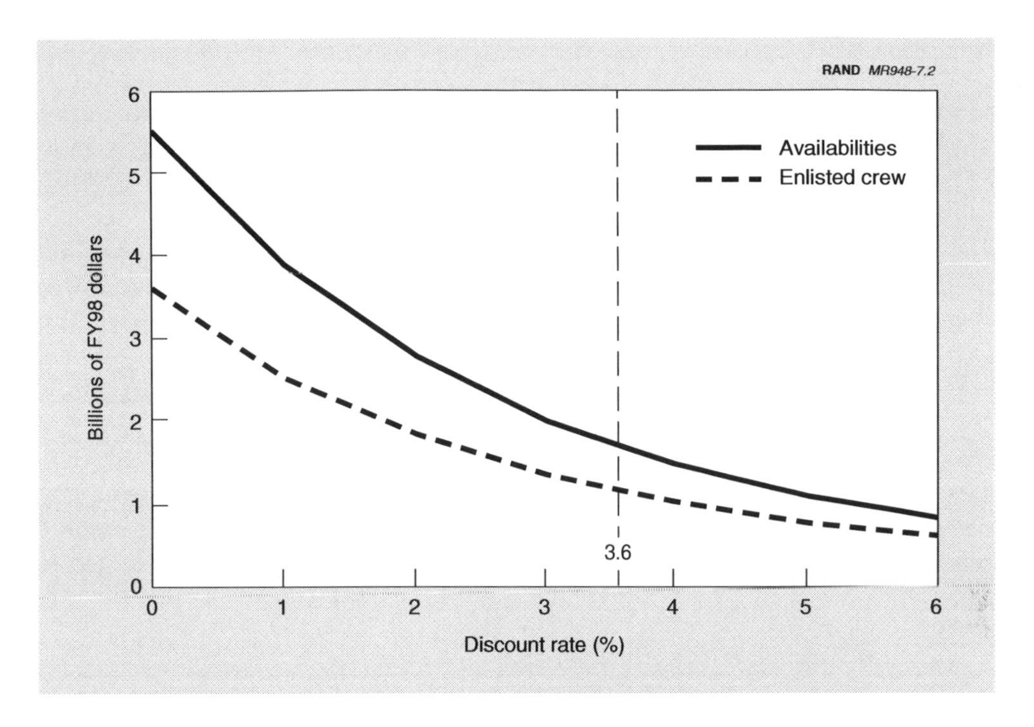

Figure 7.2—Net Present Value of CVN 77 Scheduled-Availability and Enlisted-Crew Costs, for Different Discount Rates

considering a range from 10 to over 50 percent. Using the most-conservative value of 10 percent—net of the costs of procurement, installation, and implementation associated with the new processes and technologies—the Navy should be willing to invest up to $170 million.

The NPV of the enlisted-crew costs at a 3.6-percent discount rate is $1.2 billion. A 10-percent net savings of this cost would justify an investment of up to $120 million. Combining the availability and enlisted-crew costs indicates a $290-million investment.

Estimated Investment Considering Savings for the Nimitz Class

The preceding analysis considers savings from only one ship: CVN 77. If some of the cost-reducing improvements implemented on CVN 77 could be back-fitted to the rest of the Nimitz class, the payoff would be even greater. In addition, many of the improvements—perhaps all—could be carried forward to the next class of carriers (CVX).

Analysis of potential CVX savings is beyond the scope of this study. However, we can give an indication of the payoff from a Nimitz-class backfit here. For this analysis, we assume that the other ships in the Nimitz class can be backfitted in time for operations beginning in FY09. For the backfit case, expenditures would occur between FY98 and FY09. However, we do not attempt a breakdown of expenditures by year; instead, we make the conservative assumption that all expenditures are funded in FY98.[13]

The combined scheduled-availability costs for all ships in the Nimitz class are plotted in Figure 7.3. The figure shows costs starting in FY09 and continuing to the retirement of CVN 77. The net present value of these costs is shown in Figure 7.4. The top line in the figure corresponds to the total of all costs for all ships. We cannot, however, simply assume that 10 percent of these costs could be saved, as we did for CVN 77 alone. It is not reasonable to expect that ships that are already operational will be able to take advantage of all the improvements incorporated into CVN 77. We therefore need to allow for the possibility that, for the rest of the class, the costs that could conceivably be saved—i.e., the base that is to be multiplied by some assumed percentage savings—are lower

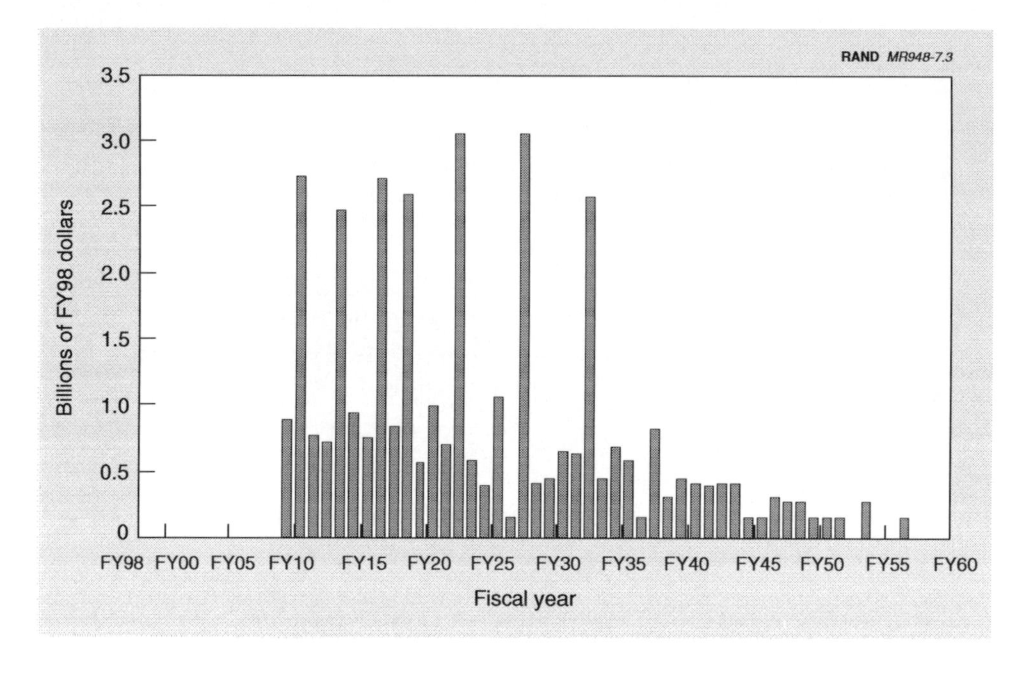

Figure 7.3—Anticipated Costs of Nimitz-Class Scheduled Availabilities, by Year

[13]To properly account for expenditures, we would, of course, need to estimate the amounts by year and then calculate the NPV of the entire stream of expenditures and savings. To the extent that R&D on new technologies and processes occurs after FY98, its NPV is smaller than that calculated here for savings, and the Navy should be willing to invest more than the amounts given in this chapter.

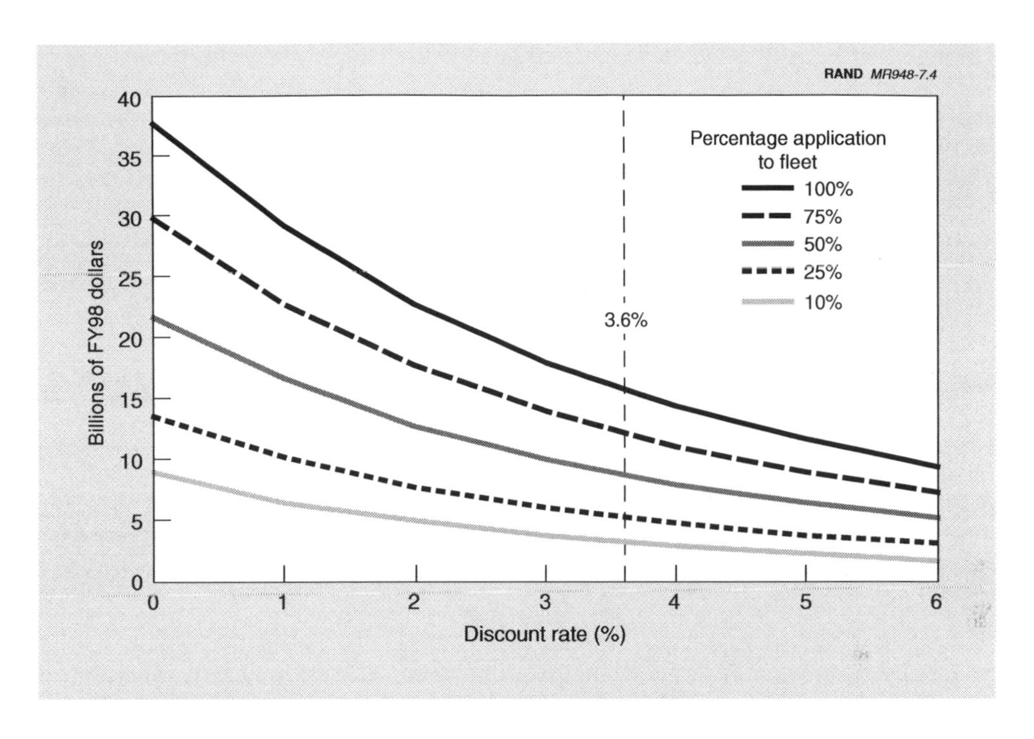

Figure 7.4—Net Present Value of Nimitz-Class Scheduled-Availability Costs, for Different Breadths of Application

than the percentage possible for CVN 77. Hence, the other lines in the figure indicate the NPV for lesser degrees of implementation of cost-savings approaches for the rest of the class. The lowest line shows the NPV O&S savings if all cost-savings measures are implemented in CVN 77 and 10 percent of them are implemented in the other ships of the class.[14] The second line from the bottom represents CVN 77 plus 25-percent application to the other ships. And so on. Even if only 10 percent of the improvements can be implemented in the rest of the class, the NPV of potentially savable costs would be about $3.1 billion at 3.6-percent discount, or $1.4 billion more than for implementation in CVN 77 alone. How much of that $3.1 billion might actually be saved? Again using the conservative 10-percent value for savings, we see that the amount would be $310 million and would justify an investment of that size.

Turning to enlisted-crew costs, we chart these costs for the Nimitz class, from FY09 to the retirement of CVN 77, in Figure 7.5 and the NPV of these costs in

[14]And if the 10 percent backfitted have a savings potential typical of any 10-percent sample drawn from the set of all improvements.

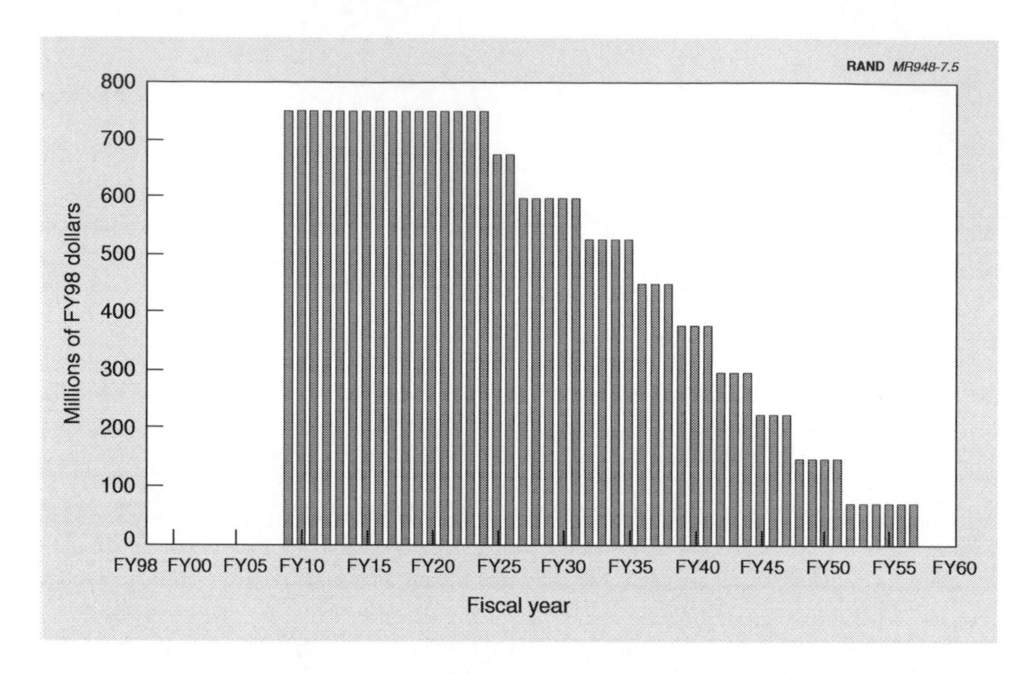

Figure 7.5—Anticipated Costs of the Enlisted Crew for the Nimitz Class, by Year

Figure 7.6, using the same format as that in Figure 7.4. If we assume that only 10 percent of the crew costs in CVN 69 through CVN 76 is a candidate for savings initiatives, in addition to 100 percent of CVN 77, then the NPV of crew costs at a 3.6-percent discount rate would be $2.0 billion, or $0.8 billion more than for CVN 77 alone. If a net 10 percent of those costs could be saved, the maximum justifiable investment would be $200 million. Combining the availability and enlisted-crew savings for CVN 77 plus a 10-percent extension to the rest of the class, we conclude that an investment of up to half a billion dollars would be justified.

SUMMARY

The Nimitz-class aircraft carriers will be a significant part of the SCN, MPN, and O&MN budgets for many years. It is important that the Navy take actions now to reduce these significant future costs. One step in this direction is to use CVN 77 as a *transition ship*, a ship in which the Nimitz-class design is modified to allow for cost-saving technologies and production processes. Our literature review and interviews suggest that the wide range of subsystems and manufacturing techniques used by European builders of commercial ships offers promise for reducing costs or improving operational availability. Our initial

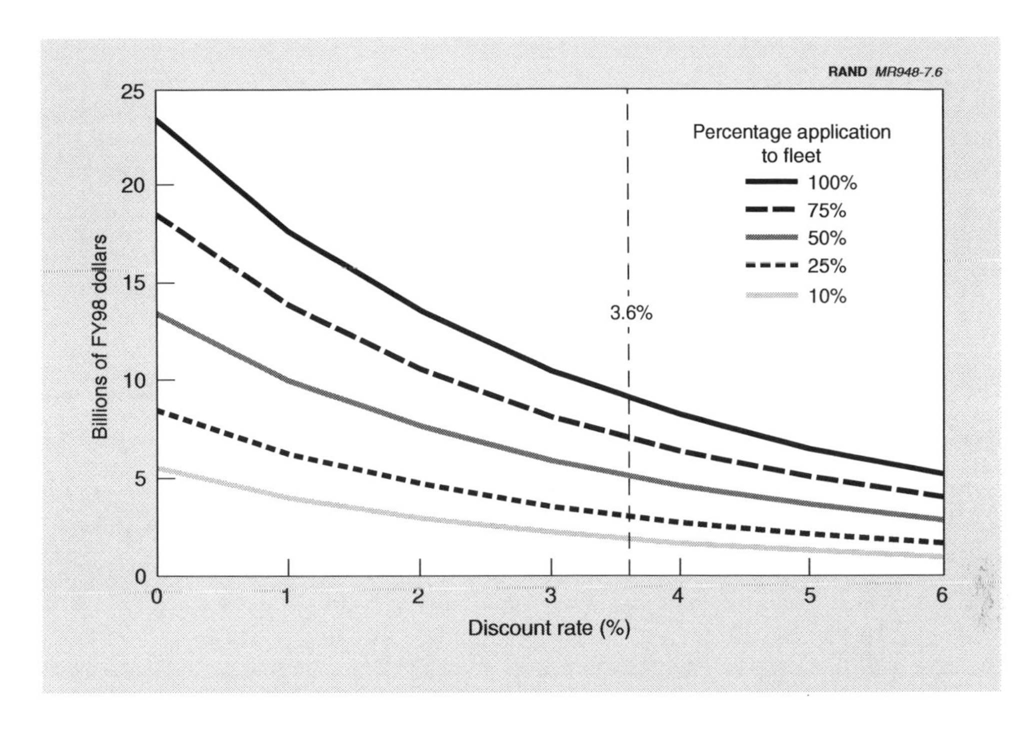

Figure 7.6—Net Present Value of Nimitz-Class Enlisted-Crew Costs, for Different Breadths of Application

analysis of the future Nimitz-class maintenance and personnel costs suggests that CVN 77 warrants R&D funding of over a quarter billion dollars to identify cost-savings technologies and production processes (see Table 7.3). That amount appears justified in view of the potential for savings on CVN 77 itself. If, through backfitting, even 10 percent of these same technologies could be applied to the other ships in the class, as much as $200 million in additional R&D funding could be justified. In arriving at that amount, we do not take credit for any projected savings in the CVX class.

Clearly, because carrier O&S costs are so large, even small-percentage reductions can save the Navy hundreds of millions of dollars. Savings on that order would appear to justify an annual R&D budget aimed at reducing the life-cycle costs of carriers and other ships in the force structure.

Table 7.3

Summary of Nimitz-Class Scheduled-Availability and Enlisted-Crew Costs and Potential Savings from Near-Term R&D

Category	Amount (FY98 $)
CVN 77 alone	
Total availability costs (NPV)	$1.7 billion
Total enlisted-crew costs (NPV)	$1.2 billion
Total of both categories (NPV)	$2.9 billion
R&D justifiable if 10% can be saved[a]	$290 million
CVN 77 + 10% backfit to rest of class	
Availability costs addressed[b] (NPV)	$3.1 billion
Enlisted-crew costs addressed[b] (NPV)	$2.0 billion
Total of both categories (NPV)	$5.1 billion
R&D justifiable if 10% can be saved[a]	$510 million

[a]That is, if gross savings minus the costs of procurement, installation, and implementation associated with the new processes and technologies is at least 10 percent of the total on the line preceding.

[b]Equals 100 percent of CVN 77 costs plus 10 percent of total costs for rest of class.

Chapter Eight

CONCLUSIONS AND RECOMMENDATIONS

Our principal conclusion regarding the aircraft carrier industrial base is this: The Newport News Shipbuilding facilities, and the supporting industrial base throughout the United States, are expected to retain the basic capabilities necessary to build large, nuclear-powered aircraft carriers into the foreseeable future, regardless of when or even whether CVN 77 is built. However, failure to start CVN 77 in the 2000–2002 time frame will inevitably lead to some decay in the quality of those capabilities and, hence, to increases in costs, schedule durations, and risks when the next carrier is started.

FINDINGS

We answer the specific research questions posed in Chapter One as follows:

1. If the current carrier force size of 12 ships is to be sustained, CVN 77 cannot be started more than a year or so beyond the currently planned date of 2002.
2. The earlier CVN 77 is started, the less it will cost. For most start dates, increasing the build period from the planned 6.5 years to 8.5 years will also reduce costs. The combination of these two effects could result in savings on the order of $400 million in shipyard labor costs.
3. Timing of the CVN 77 start should not greatly affect the survival of vendors supplying nuclear or nonnuclear components to the shipyard. However, modest cost savings should accrue from ordering some contractor-furnished equipment two years or more in advance of the beginning of work in the shipyard.
4. Many cost-saving production processes and technologies are being implemented by builders of large, complex commercial ships but are not employed in building ships funded by the U.S. Navy. By adopting these technologies and processes, the U.S. Navy has a savings potential on CVN 77 alone, conservatively estimated, of over a quarter billion dollars.

RECOMMENDATIONS

Our findings lead to the following recommendations:

- Begin ship fabrication before 2002. The potential for savings here is substantial—in the hundreds of millions of dollars.
- Order contractor-furnished equipment in advance of shipyard start. Doing so should permit additional savings in the tens of millions of dollars.
- Invest at least a quarter billion dollars in research and development directed toward adapting production processes and application-engineering improvements that could reduce the cost of carrier construction, operation and maintenance, and manning. The costs involved in building and operating carriers are, in fact, so large that the Navy should consider establishing a stable annual R&D funding level for these ships.

Appendix A

CARRIER DATA[1]

This appendix contains two tables of carrier data. Table A.1 summarizes the production and disposition history of all U.S. aircraft carriers to date. Table A.2 provides dimensions, crew sizes, and propulsion and power specifications for most carriers built from *Forrestal* on.

[1]Data in the appendix were taken from various volumes of *Jane's Fighting Ships* (London: Jane's Information Group), *The World Aircraft Carrier Lists* (available online at http://www.uss-salem.org/navhist/carriers/) and various communications from Newport News Shipbuilding.

Table A.1

Production and Disposition History of All U.S. Aircraft Carriers

Hull No.	Name	Class	Builder[a]	Laid Down	Launched	Commis-sioned	Decom. (D), Sunk (S), Can. (C)	Date	Stricken[b]	Notes[c]
CV 1	*Langley*[d]	Langley	MINSY/ NNSY	18-Oct-11	24-Aug-12	7-Apr-13	S	27-Feb-42		ex AC 3; to AV 3
CV 2	*Lexington*[e]	Lexington	Beth FR	8-Jan-21	3-Oct-25	14-Dec-27	S	8-May-42		ex CC 1
CV 3	*Saratoga*[f]	Lexington	NYSB	25-Sep-20	7-Apr-25	16-Nov-27	S	25-Jul-46		ex CC 3
CV 4	*Ranger*	Ranger	NNSB	26-Sep-31	24-Feb-33	4-Jun-34	D	18-Oct-46	29-Oct-46	
CV 5	*Yorktown*	Yorktown	NNS	21-May-34	4-Apr-36	30-Sep-37	S	7-Jun-42		
CV 6	*Enterprise*	Yorktown	NNS	16-Jul-34	3-Oct-36	12-May-38	D	17-Feb-47	2-Oct-56	
CV 7	*Wasp*	Wasp	Beth FR	1-Apr-36	4-Apr-39	25-Apr-40	S	15-Sep-42		
CV 8	*Hornet*	Yorktown	NNS	25-Sep-39	14-Dec-40	25-Oct-41	S	26-Oct-42		
CV 9	*Essex*	Essex	NNS	28-Apr-41	31-Jul-42	31-Dec-42	D	30-Jun-69	1-Jun-73	
CV 10	*Yorktown*	Essex	NNS	1-Dec-41	21-Jan-43	15-May-43	D	27-Jun-70	1-Jun-73	
CV 11	*Intrepid*	Essex	NNS	1-Dec-41	26-Apr-43	16-Aug-43	D	15-Mar-74	23-Feb-82	
CV 12	*Hornet*[g]	Essex	NNS	3-Aug-42	29-Aug-43	29-Nov-43	D	26-Jun-70	25-Jul-89	
CV 13	*Franklin*	Essex	NNS	7-Dec-42	14-Oct-43	31-Jan-44	D	17-Feb-47	1-Oct-64	to AVT 8
CV 14	*Ticonderoga*	Ticonderoga[h]	NNS	1-Feb-43	7-Feb-44	10-Oct-44	D	16-Nov-73	16-Nov-73	
CV 15	*Randolph*	Ticonderoga	NNS	10-May-43	28-Jun-44	9-Oct-44	D	13-Feb-69	1-Jun-73	
CV 16	*Lexington*	Essex	Beth Q	15-Jul-41	26-Sep-42	17-Feb-43	D	26-Nov-91	30-Nov-91	to AVT 16
CV 17	*Bunker Hill*	Essex	Beth Q	15-Sep-41	7-Dec-42	24-May-43	D	Jan-47	1-Nov-66	to AVT 9
CV 18	*Wasp*	Essex	Beth Q	18-Mar-42	17-Aug-43	24-Nov-43	D	1-Jul-72	1-Jul-72	
CV 19	*Hancock*	Ticonderoga	Beth Q	26-Jan-43	24-Jan-44	15-Apr-44	D	30-Jan-76	31-Jan-76	
CV 20	*Bennington*	Essex	NYNSY	15-Dec-42	26-Feb-44	6-Aug-44	D	15-Jan-70	20-Sep-89	
CV 21	*Boxer*	Ticonderoga	NNS	13-Sep-43	14-Dec-44	16-Apr-45	D	1-Dec-69	1-Dec-69	to LPH 4
CVL 22	*Independence*	Independence[i]	NYSB	1-May-41	22-Aug-42	14-Jan-43	D	Aug-46	27-Feb-51	ex CL 59
CVL 23	*Princeton*	Independence	NYSB	6-Feb-41	18-Oct-42	25-Feb-43	S	24-Oct-44		ex CL 61
CVL 24	*Belleau Wood*	Independence	NYSB	11-Aug-41	6-Dec-42	31-Mar-43	D	13-Jan-47	1-Oct-60	ex CL 76; to France
CVL 25	*Cowpens*	Independence	NYSB	17-Nov-41	17-Jan-43	28-May-43	D	13-Jan-47	1-Nov-59	ex CL 77; to AVT 1
CVL 26	*Monterey*	Independence	NYSB	29-Dec-41	28-Feb-43	17-Jun-43	D	16-Jan-56	1-Jun-70	ex CL 78; to AVT 2

Table A.1—continued

Hull No.	Name	Class	Builder[a]	Laid Down	Launched	Commis-sioned	Decom. (D), Sunk (S), Can. (C)	Date	Stricken[b]	Notes[c]
CVL 27	*Langley*	Independence	NYSB	11-Apr-42	22-May-43	31-Aug-43	D	11-Feb-47	20-Mar-63	ex CL 85; to France
CVL 28	*Cabot*	Independence	NYSB	16-Aug-42	4-Apr-43	24-Jul-43	D	21-Jan-55	1-Aug-72	ex CL 79; to AVT 3; to Spain
CVL 29	*Bataan*	Independence	NYSB	31-Aug-42	1-Aug-43	17-Nov-43	D	9-Apr-54	1-Sep-59	ex CL 99; to AVT 4
CVL 30	*San Jacinto*	Independence	NYSB	26-Oct-42	29-Sep-43	15-Dec-43	D	1-Mar-47	1-Jun-70	ex CL 100; to AVT 5
CV 31	*Bon Homme Richard*	Essex	NYNSY	1-Feb-43	29-Apr-44	26-Nov-44	D	2-Jul-71	20-Sep-89	
CV 32	*Leyte*	Ticonderoga	NNS	21-Feb-44	23-Aug-45	11-Apr-46	D	15-May-59	1-Jun-69	to AVT 10
CV 33	*Kearsarge*	Ticonderoga	NYNSY	1-Mar-44	5-May-45	2-Mar-46	D	13-Feb-70	1-May-73	
CV 34	*Oriskany*	Ticonderoga	NYNSY	1-May-44	13-Oct-45	25-Sep-50	D	15-May-76	25-Jul-89	
CV 35	*Reprisal*	Ticonderoga	NYNSY	1-Jul-44	1946		—[j]			
CV 36	*Antietam*	Ticonderoga	PhNSY	15-Mar-43	20-Aug-44	28-Jan-45	D	8-May-63	1-May-73	
CV 37	*Princeton*	Ticonderoga	PhNSY	14-Sep-43	8-Jul-45	18-Nov-45	D	30-Jan-70	30-Jan-70	to LPH 5
CV 38	*Shangri-La*	Ticonderoga	NNSY	15-Jan-43	24-Feb-44	15-Sep-44	D	30-Jun-71	15-Jul-82	
CV 39	*Lake Champlain*	Ticonderoga	NNSY	15-Mar-43	2-Nov-44	3-Jun-45	D	2-May-66	1-Dec-69	
CV 40	*Tarawa*	Ticonderoga	NNSY	1-Mar-44	12-May-45	8-Dec-45	D	13-May-60	1-Jan-67	to AVT 12
CVB 41	*Midway*	Midway	NNS	27-Oct-43	20-Mar-45	10-Sep-45	D	11-Apr-92		
CVB 42	*Franklin D. Roosevelt*	Midway	NYNSY	1-Dec-43	29-Apr-45	27-Oct-45	D	1-Oct-77	1-Oct-77	
CVB 43	*Coral Sea*	Midway	NNS	10-Jul-44	2-Apr-46	1-Oct-47	D	30-Apr-91	30-Apr-91	
CVB 44	(none)	Midway	NNS				C	11-Jan-43	(no bldg dock available at NNS)	
CV 45	*Valley Forge*	Essex	PhNSY	7-Sep-44	18-Nov-45	3-Nov-46	D	15-Jan-70	15-Jan-70	to LPH 8
CV 46	*Iwo Jima*	Essex	NNS	29-Jan-45			C	11-Aug-45		
CV 47	*Philippine Sea*	Essex	Beth Q	19-Aug-44	5-Sep-45	11-May-46	D	28-Dec-58	1-Dec-69	to AVT 11
CVL 48	*Saipan*	Saipan	NYSB	10-Jul-44	8-Jul-45	14-Jul-46	D	14-Jan-70	15-Aug-75	to AVT 5; CC 3; AGMR 2
CVL 49	*Wright*	Saipan	NYSB	21-Aug-44	1-Sep-45	9-Feb-47	D	27-May-70	1-Dec-77	to AVT 7; to CC 2
CV 50	(none)	Essex	Beth Q				C	28-Mar-45		

Table A.1—continued

Hull No.	Name	Class	Builder[a]	Laid Down	Launched	Commis-sioned	Decom. (D), Sunk (S), Can. (C)	Date	Stricken[b]	Notes[c]
CV 51	(none)	Essex	NYNSY				C	28-Mar-45		
CV 52	(none)	Essex	NYNSY				C	28-Mar-45		
CV 53	(none)	Essex	PhNSY				C	28-Mar-45		
CV 54	(none)	Essex	NNSY				C	28-Mar-45		
CV 55	(none)	Essex	NNSY				C	28-Mar-45		
CVB 56	(none)	Midway	NNS				C	28-Mar-45		
CVB 57	(none)	Midway	NNS				C	28-Mar-45		
CVA 58	*United States*	United States	NNS	18-Apr-49			C	23-Apr-49		scrapped
CVA 59	*Forrestal*	Forrestal	NNS	14-Jul-52	11-Dec-54	1-Oct-55	D	10-Sep-93	10-Sep-93	to AVT 59
CVA 60	*Saratoga*	Forrestal	NYNSY	16-Dec-52	8-Oct-55	14-Apr-56	D	30-Sep-94	30-Sep-94	
CVA 61	*Ranger*	Forrestal	NNS	2-Aug-54	29-Sep-56	10-Aug-57	D	10-Jul-93		
CVA 62	*Independence*	Forrestal	NYNSY	1-Jul-55	6-Jun-58	10-Jan-59	D	1998		
CVA 63	*Kitty Hawk*	Kitty Hawk	NYSB	27-Dec-56	21-May-60	29-Apr-61	D	2002		
CVA 64	*Constellation*	Kitty Hawk	NYNSY	14-Sep-57	8-Oct-60	27-Oct-61	D	2008		
CVA 65	*Enterprise*	Enterprise	NNS	4-Feb-58	24-Sep-60	25-Nov-61	D	2014		
CVA 66	*America*	Kitty Hawk	NNS	9-Jan-61	1-Feb-64	23-Jan-65	D	9-Aug-96		
CVA 67	*John F. Kennedy*	Kennedy	NNS	22-Oct-64	27-May-67	7-Sep-68	D	2011		
CVA 68	*Nimitz*	Nimitz	NNS	22-Jun-68	13-May-72	3-May-75	D	2022		
CVA 69	*Dwight D. Eisenhower*	Nimitz	NNS	15-Aug-70	11-Oct-75	18-Oct-77	D	2022		
CVA 70	*Carl Vinson*	Nimitz	NNS	11-Oct-75	15-Mar-80	13-Mar-82	D	2027		
CVN 71	*Theodore Roosevelt*	Nimitz	NNS	13-Oct-81	27-Oct-84	25-Oct-86	D	2031		
CVN 72	*Abraham Lincoln*	Nimitz	NNS	3-Nov-84	13-Feb-88	11-Nov-89	D	2034		
CVN 73	*George Washington*	Nimitz	NNS	25-Aug-86	21-Jul-90	4-Jul-92	D	2037		
CVN 74	*John C. Stennis*	Nimitz	NNS	13-Mar-91	13-Nov-93	9-Dec-95	D	2040		
CVN 75	*Harry S. Truman*	Nimitz	NNS	29-Nov-93	Sep-96	12-Jun-98	D	2043		
CVN 76	*Ronald Reagan*	Nimitz	NNS	Feb-98	Mar-00	Dec-02	D	2047		

Table A.1—continued

Hull No.	Name	Class	Builder[a]	Laid Down	Launched	Commis-sioned	Decom. (D), Sunk (S), Can. (C)	Date	Stricken[b]	Notes[c]
CVN 77	(none)	Nimitz	NNS	Feb-05	Mar-07	Dec-09				
CVX 78	(none)	unknown		Feb-09	Mar-11	Dec-13				

[a]MINSY = Mare Island Naval Shipyard, NNSY = Norfolk Naval Shipyard, Beth FR = Bethlehem Forge River, NYSB = New York Shipbuilding, NNS = Newport News Shipbuilding, Beth Q = Bethlehem Quincy, NYNSY = New York Naval Shipyard, PhNSY = Philadelphia Naval Shipyard.

[b]*Stricken* means that a vessel has been formally determined by the Secretary of the Navy to have no further value to the nation and is awaiting disposition or has been disposed of.

[c]Notes give previous ("ex") and subsequent ("to") hull designations or country sold to.

[d]Built as collier *Jupiter* at MINSY. Converted to CV at NNSY beginning 24 Mar 20; finished 20 Mar 22; commissioned 20 Mar 22.

[e]Laid down as battle cruiser *Constitution*; suspended 8 Feb 1922; launched as CV, 3 Oct 1925; commissioned 14 Dec 27.

[f]Laid down as battle cruiser; suspended 8 Feb 1922; launched as CV, 7 Apr 1925; commissioned 16 Nov 27.

[g]Originally laid down early 1942. Work stopped and keel lifted out of dock to allow LST construction.

[h]Modified Essex class.

[i]All Independence-class ships were light cruiser hulls.

[j]Scrapped August 1949 (used for explosives tests).

Classification:

AC	=	collier.
AGMR	=	a major communications-relay ship.
AV	=	a seaplane tender.
AVT	=	auxiliary aircraft transport.
CC	=	cruiser, battle, or command ship.
CL	=	cruiser, light.
CV	=	established 1922 when *Langley* was converted.
CVB	=	established 15 July 1943 for armored-deck ships; reclassified to CVA, 1 Oct 1952.
CVL	=	established 15 July 1943 for light carriers; last stricken 1 June 1979.
CVA	=	established 1 Oct 1952 for offensive-role carriers; reclassified to CV, 30 June 1975.
CVS	=	established 8 Aug 1953 for anti-submarine warfare (ASW) ships; last stricken 20 Sept 1989.
CVT	=	established 1 Jan 1969 for training ships; reclassified to AVT, 16 July 1978.
CVAN	=	established 15 Nov 1957 for nuclear-powered ships; reclassified to CVN, 30 June 1975.
CVN	=	established 30 June 1975.
LPH	=	landing platform–helicopter.
LST	=	landing ship–tank.

Table A.2

Specifications for Aircraft Carriers from *Forrestal* On

a. Size Specifications

		Displacement		Dimensions (ft)			Number of Personnel					
							Crew		Aircrew		Flag	
Hull No.	Name	(std/lt)	(loaded)	length	beam	draft	(officer)	(enlisted)	(officer)	(enlisted)	(officer)	(enlisted)
CV 59	*Forrestal*	59,060	79,250	1,086	130	37	154	2,746	329	1,950	25	45
CV 60	*Saratoga*	59,060	80,383	1,063	130	37	154	2,746	329	1,950	25	45
CV 62	*Independence*	60,000	80,643	1,071	130	37	154	2,746	329	1,950	25	45
CV 63	*Kitty Hawk*	60,100	81,123	1,062.5	130	37.4	155	2,775	320	2,160	25	45
CV 64	*Constellation*	60,100	81,773	1,072.5	130	37.4	155	2,775	320	2,160	25	45
CVN 65	*Enterprise*	73,502	93,970	1123	133	39	171	3,044	358	2,122	25	45
CV 66	*America*	60,300	79,724	1,047.5	130	37.4	155	2,775	320	2,160	25	45
CV 67	*John F. Kennedy*	61,000	80,941	1,052	130	37.4	155	2,775	320	2,160	25	45
CVN 68	*Nimitz*	72,916	91,487	1,092	134	37	203	2,981	366	2,434	25	45
CVN 69	*Dwight D. Eisenhower*	72,916	91,487	1,092	134	37	203	2,981	366	2,434	25	45
CVN 70	*Carl Vinson*	72,916	91,487	1,092	134	37	203	2,981	366	2,434	25	45
CVN 71	*Theodore Roosevelt*	73,973	96,386	1,092	134	38.7	203	2,981	366	2,434	25	46
CVN 72	*Abraham Lincoln*		102,000	1,092	134	39	203	2,981	366	2,434	25	45
CVN 73	*George Washington*		102,000	1,092	134	39	203	2,981	366	2,434	25	45
CVN 74	*John C. Stennis*		102,000	1,092	134	39	203	2,981	366	2,434	25	45

lt = long tons (British measure of volume); std = standard.

Table A.2

Specifications for Aircraft Carriers from *Forrestal* On

b. Propulsion Specifications

			Standard Turbines		Emergency Turbines			
Hull No.	Name	Main Propulsion	(No.)	(horsepower)	(No.)	(horsepower)	Shafts	Speed (kt)
CV 59	*Forrestal*	8 Babcock&Wilcox boilers	4	260,000			4	33
CV 60	*Saratoga*	8 Babcock&Wilcox boilers	4	280,000			4	33
CV 62	*Independence*	8 Babcock&Wilcox boilers	4	280,000			4	33
CV 63	*Kitty Hawk*	8 Foster-Wheeler boilers	4	280,000			4	32
CV 64	*Constellation*	8 Foster-Wheeler boilers	4	280,000			4	32
CVN 65	*Enterprise*	8 Westinghouse PWR A2W	4	280,000	4 diesel	10,720	4	33
CV 66	*America*	8 Foster-Wheeler boilers	4	280,000			4	32
CV 67	*John F. Kennedy*	8 Foster-Wheeler boilers	4	280,000			4	32
CVN 68	*Nimitz*	2 GE PWR A4W/A1G	4	260,000	4 diesel	10,720	4	30+
CVN 69	*Dwight D. Eisenhower*	2 GE PWR A4W/A1G	4	260,000	4 diesel	10,720	4	30+
CVN 70	*Carl Vinson*	2 GE PWR A4W/A1G	4	260,000	4 diesel	10,720	4	30+
CVN 71	*Theodore Roosevelt*	2 GE PWR A4W/A1G	4	260,000	4 diesel	10,720	4	30+
CVN 72	*Abraham Lincoln*	2 GE PWR A4W/A1G	4	260,000	4 diesel	10,720	4	30+
CVN 73	*George Washington*	2 GE PWR A4W/A1G	4	260,000	4 diesel	10,720	4	30+
CVN 74	*John C. Stennis*	2 GE PWR A4W/A1G	4	260,000	4 diesel	10,720	4	30+

PWR = pressurized water reactors (nuclear).

Appendix B

CRISIS-RESPONSE DATA, 1950–1996

In Chapter Two, we offered some information on crisis response by aircraft carriers. Here, we elaborate on that information and place it in the context of U.S. military crisis response in general. Any such compilation is necessarily subjective. Thus, for the data to be meaningful, the criteria for inclusion of events as "crises" and sources from which the data were drawn must be fully understood. Much work on military operations in other than major conflicts has already been done and is ongoing, so we considered it prudent to use data from such studies. They are

- *The Use of Naval Forces in the Post-War Era: U.S. Navy and U.S. Marine Corps Crisis Response Activity, 1946–1990,* Adam B. Siegel, Alexandria, Va.: Center for Naval Analyses, CRM-90-246, February 1991.
- *Answering the 9-1-1 Call: U.S. Military and Naval Crisis Response Activity, 1977–91,* Thomas P.M. Barnett, Linda D. Lancaster, Alexandria, Va.: Center for Naval Analyses, ADB173802, August 1992.
- *Preparing the U.S. Air Force for Military Operations Other Than War,* Alan Vick, David T. Orletsky, Abram N. Shulsky, John Stillion, Santa Monica, Calif.: RAND, MR-842-AF, 1997.

Additional data on U.S. Army (USA) operations were obtained from the Army Center for Military History.

The four separate databases from these sources were combined into the single database in Table B.1. We excluded the following operations:

- The three major military conflicts during the period June 1950–September 1996 (Korean War, Vietnam War, 1991 Gulf War). However, pre- and post-war events related to the conflicts were included.
- Events that took place within the United States.

- Purely humanitarian or disaster-relief operations. However, some operations considered to be primarily humanitarian but that also had a broader geopolitical rationale were included (e.g., Operation Provide Comfort).
- Alerts or other actions not involving force movements.
- Intelligence operations.
- Routine operations in support of U.S. diplomacy.
- Law enforcement and counternarcotics operations.
- Routine training and assistance to allies, airlift operations, and exercises.

A few other points will be of help in interpreting the results:

- Only the starting date of the crises in the table has been noted, not the duration. In some cases, the duration is clear and easily defined; in others, it is difficult to determine.
- Some actions that begin as a response to a crisis eventually become longstanding, routine operations. For example, deployments to the Indian Ocean/Persian Gulf were initially in response to the Iranian revolution and Soviet invasion of Afghanistan in 1979; since then, this area has remained a regular theater of U.S. operations. NATO combat air patrols over Bosnia-Herzegovina (Operation Deny Flight) are another example. In such cases, the initial response to the original crisis is counted in the database; the continuing presence of U.S. forces is not reflected. Conversely, some short and easily identified missions relating to the same situation could be combined into one larger operation—a selectivity that obviously affects both the final total of crisis responses and individual-service participation. However, every meaningful instance of service participation has been included in order to present as unbiased a picture as possible.
- Finally, it must be noted that U.S. Marine Corps forces and actions are counted and included in Navy forces and actions. This is not intended to diminish Marine Corps participation in such actions, which is continually—and sometimes inordinately—high. Rather, it is merely a reflection of the fact that Marine Corps activity is not relevant to this study, but that of naval forces as a whole and aircraft carriers in particular is.

Table B.1

Chronological List of U.S. Military Crisis Responses, 1950–1996

No.	Operation/Event/Location	Date Begun	USN Involved?	If So, No. of CVs Used	USAF Involved?	U.S. Army Involved?
1.	Korean War; Formosa Straits	Jun 50	Y	1	N	N
2.	Korean War; Security in Europe	Jul 50	Y	2	Y	Y
3.	Lebanon	Aug 50	Y	2	N	N
4.	Security of Yugoslavia	Mar 51	Y	2	N	N
5.	China-Taiwan Conflict	Feb 53	Y	1	N	N
6.	Dien Bien Phu	Mar 54	Y	2	Y	N
7.	Honduras-Guatemala	May 54	Y	1	N	N
8.	People's Republic of China (PRC) Shootdown	Jul 54	Y	2	N	N
9.	Vietnam Evacuations	Aug 54	Y	0	N	N
10.	Honduran Elections	Oct 54	Y	0	N	N
11.	Accord on Trieste	Oct 54	Y	0	N	Y
12.	Tachen Islands	Feb 55	Y	6	Y	N
13.	Red Sea Patrols	Feb 56	Y	0	N	N
14.	Jordan	Mar 56	Y	2	N	N
15.	Pre-Suez	Aug 56	Y	2	N	N
16.	Suez War	Oct 56	Y	3	Y	N
17.	Port Lyautey	Nov 56	Y	0	N	N
18.	Post-Suez	Nov 56	Y	8	N	N
19.	Cuban Civil War	Dec 56	Y	1	N	N
20.	Red Sea Patrols	Feb 57	Y	0	N	N
21.	Jordan Unrest	Apr 57	Y	2	N	N
22.	Haiti	Jun 57	Y	0	N	N
23.	PRC–Republic of China (ROC) Tension	Jul 57	Y	3	N	N
24.	Syria	Aug 57	Y	4	Y	N
25.	Indonesia	Dec 57	Y	2	N	N
26.	Venezuelan Revolution	Jan 58	Y	0	N	N
27.	Laos	Mar 58	N	0	Y	N
28.	Venezuela	May 58	Y	0	N	Y
29.	Lebanon	May 58	Y	3	Y	N
30.	Lebanon	Jul 58	Y	3	Y	Y
31.	Jordan-Iraq	Jul 58	Y	0	N	N
32.	Quemoy	Aug 58	Y	6	Y	Y
33.	Panama	Apr 59	Y	0	N	N
34.	Berlin Crisis	May 59	Y	2	Y	Y
35.	Laos	Jul 59	Y	1	Y	Y
36.	PRC-ROC	Jul 59	Y	2	N	N
37.	Panama	Aug 59	Y	0	N	N
38.	Congo	Jul 60	Y	1	Y	Y
39.	Guatemala	Nov 60	Y	2	N	N
40.	Laos	Jan 61	Y	3	Y	Y
41.	SS *Santa Maria*	Jan 61	Y	0	N	N
42.	Gulf of Guinea–Congo	Feb 61	Y	0	N	N
43.	Laos	Mar 61	Y	3	Y	N
44.	SS *Western Union*	Mar 61	Y	0	N	N
45.	Bay of Pigs	Apr 61	Y	2	Y	N
46.	Dominican Republic	May 61	Y	3	Y	Y

Table B.1—continued

No.	Operation/Event/Location	Date Begun	USN Involved?	If So, No. of CVs Used	USAF Involved?	U.S. Army Involved?
47.	Zanzibar	Jun 61	Y	0	N	N
48.	Kuwait	Jul 61	Y	0	N	N
49.	Berlin Crisis	Jul 61	Y	3	Y	Y
50.	Taiwan	Aug 61	N	0	Y	N
51.	Dominican Republic	Nov 61	Y	1	N	N
52.	Thailand	Nov 61	N	0	Y	N
53.	South Vietnam	Dec 61	Y	0	Y	Y
54.	Dominican Republic	Jan 62	Y	0	N	N
55.	Guatemala Riots	Mar 62	Y	1	Y	N
56.	South Vietnam	Apr 62	Y	0	N	N
57.	Laos/Thailand	May 62	Y	2	Y	Y
58.	Guantanamo	Jul 62	Y	0	N	N
59.	Haiti Civil Disorder	Aug 62	Y	1	N	N
60.	Yemen	Sep 62	Y	0	N	N
61.	Cuban Missile Crisis	Sep 62	Y	8	Y	Y
62.	Sino-Indian War	Nov 62	Y	1	Y	N
63.	SS *Anzoatequi*	Feb 63	Y	0	N	N
64.	Laos	Apr 63	Y	2	N	N
65.	Haitian Unrest	Apr 63	Y	1	N	N
66.	Haiti Civil War	Aug 63	Y	1	N	N
67.	Vietnam Civil Disorder	Aug 63	Y	2	N	N
68.	PRC-ROC	Sep 63	Y	1	N	N
69.	Dominican Republic	Sep 63	Y	0	N	N
70.	Indonesia-Malaysia	Oct 63	Y	1	N	N
71.	Zanzibar	Jan 64	Y	0	N	N
72.	Tanganyika	Jan 64	Y	0	N	N
73.	Caribbean Surveillance	Jan 64	Y	0	N	N
74.	Panama	Jan 64	Y	0	Y	Y
75.	Venezuela	Jan 64	Y	0	N	N
76.	Cyprus	Jan 64	Y	1	Y	N
77.	Peru	Mar 64	N	0	Y	N
78.	Brazil	Mar 64	Y	1	N	N
79.	Laos	Apr 64	Y	2	Y	N
80.	Guantanamo	May 64	Y	0	N	N
81.	Panama	May 64	Y	0	N	N
82.	Dominican Republic	Jul 64	Y	0	N	N
83.	Gulf of Tonkin	Aug 64	Y	2	N	N
84.	Haiti	Aug 64	Y	0	N	N
85.	Congo Noncombatant Evacuation Order (NEO)	Aug 64	N	0	Y	N
86.	Congo NEO	Nov 64	N	0	Y	N
87.	Panama	Jan 65	Y	0	N	N
88.	Tanzania	Jan 65	Y	0	N	N
89.	Venezuela-Colombia	Jan 65	Y	0	N	N
90.	British Guiana	Apr 65	Y	0	N	N
91.	Dominican Republic	Apr 65	Y	2	Y	Y
92.	Yemen	Jul 65	Y	0	N	N
93.	Ethiopia Hostage Rescue	Jul 65	N	0	Y	N
94.	Cyprus	Aug 65	Y	1	Y	N
95.	Indo-Pakistani War	Sep 65	Y	0	Y	N

Table B.1—continued

No.	Operation/Event/Location	Date Begun	USN Involved?	If So, No. of CVs Used	USAF Involved?	U.S. Army Involved?
96.	Indonesia	Oct 65	Y	0	N	N
97.	Greek Coup	Apr 67	Y	1	N	N
98.	Six-Day War	Jun 67	Y	2	Y	Y
99.	Congo	Jul 67	N	0	Y	N
100.	Destroyer *Eilat* Sinking	Oct 67	Y	2	N	N
101.	Cyprus	Nov 67	Y	1	N	N
102.	USS *Pueblo*	Jan 68	Y	3	Y	N
103.	EC-121 Shootdown	Apr 69	Y	4	Y	Y
104.	Curaçao Civil Unrest	May 69	Y	0	N	N
105.	Lebanon-Libya Operations (Ops)	Oct 69	Y	2	N	N
106.	Trinidad	Apr 70	Y	0	Y	N
107.	Jordan	Jun 70	Y	1	Y	N
108.	Jordan	Sep 70	Y	3	Y	Y
109.	Haiti Succession	Apr 71	Y	0	N	N
110.	Indo-Pakistani War	Dec 71	Y	1	N	N
111.	Bahama Lines	Dec 71	Y	0	N	N
112.	Taiwan Air Defense	Nov 72	N	0	Y	N
113.	Lebanon	May 73	Y	2	N	N
114.	Middle East War	Oct 73	Y	3	Y	Y
115.	Middle East Force	Oct 73	Y	0	N	N
116.	Oil Embargo—Indian Ocean Ops	Oct 73	Y	1	N	N
117.	Cyprus	Jul 74	Y	2	Y	Y
118.	Cyprus Unrest	Jan 75	Y	1	N	N
119.	Ethiopia	Feb 75	Y	0	N	N
120.	Eagle Pull, Cambodia	Feb 75	Y	1	Y	Y
121.	Vietnam Nuclear Transport	Mar 75	N	0	Y	N
122.	Frequent Wind, Vietnam	Apr 75	Y	4	Y	N
123.	SS *Mayaguez*	May 75	Y	2	Y	N
124.	Lebanon	Aug 75	Y	1	N	N
125.	Polisario Rebels	Jan 76	Y	0	N	N
126.	Lebanon NEO	Jun 76	N	0	Y	N
127.	Tunisia	Jul 76	Y	0	N	N
128.	Kenya-Uganda	Jul 76	Y	1	N	N
129.	Korean Tree Incident	Aug 76	Y	1	Y	Y
130.	Uganda	Feb 77	Y	1	N	N
131.	Ogaden War	Feb 78	Y	1	Y	N
132.	Zaire	May 78	N	0	Y	N
133.	Sea of Okhotsk	Jun 78	Y	0	N	N
134.	Afghanistan	Jul 78	Y	1	N	N
135.	Nicaragua	Sep 78	Y	0	Y	N
136.	Israel-Lebanon	Sep 78	N	0	Y	N
137.	Iranian Revolution	Dec 78	Y	1	Y	N
138.	Saudi Arabia	Mar 79	N	0	Y	N
139.	China-Vietnam	Feb 79	Y	1	N	N
140.	Yemen	Mar 79	Y	1	Y	N
141.	Nicaraguan Revolution	Jul 79	Y	0	Y	N
142.	Soviet Troops in Cuba	Oct 79	Y	1	Y	N
143.	Afghan/Iran Hostages	Oct 79	Y	2	Y	Y

Table B.1—continued

No.	Operation/Event/Location	Date Begun	USN Involved?	If So, No. of CVs Used	USAF Involved?	U.S. Army Involved?
144.	Park-Chung Hee	Oct 79	Y	1	Y	Y
145.	Bolivia	Nov 79	N	0	Y	N
146.	Zimbabwe	Dec 79	N	0	Y	N
147.	Iran Hostage Rescue	Apr 80	Y	1	Y	Y
148.	Korea	May 80	Y	1	Y	Y
149.	Thailand	Jun 80	N	0	Y	N
150.	Iran-Iraq War	Sep 80	Y	2	Y	N
151.	Poland	Dec 80	Y	0	Y	N
152.	Saudi Arabia	Jan 81	N	0	Y	N
153.	El Salvador	Jan 81	N	0	Y	N
154.	Morocco	Jan 81	Y	0	N	N
155.	Liberia	Apr 81	Y	0	N	Y
156.	Syria	May 81	Y	2	N	N
157.	Sadat visits Sudan	May 81	N	0	Y	N
158.	Gambia NEO	Jul 81	N	0	Y	N
159.	Libya	Aug 81	Y	2	N	N
160.	Sadat-Sudan	Oct 81	Y	1	Y	N
161.	Central America	Oct 81	Y	2	N	N
162.	Korea	Dec 81	N	0	Y	N
163.	El Salvador	Mar 82	N	0	Y	N
164.	Falklands	May 82	N	0	Y	N
165.	Israeli Invasion	Jun 82	Y	1	Y	N
166.	Somalia	Jul 82	N	0	Y	N
167.	Lebanon Peacekeeping Force	Aug 82	Y	2	N	N
168.	Palestinian Massacre	Sep 82	Y	2	N	N
169.	Libya-Sudan	Feb 83	Y	1	Y	N
170.	Thailand-Burma	Apr 83	N	0	Y	N
171.	Honduras	Jun 83	Y	1	Y	Y
172.	Libya-Chad	Aug 83	Y	1	Y	N
173.	Marine Barracks Bomb	Aug 83	Y	2	Y	Y
174.	Korea Airlines 007	Sep 83	Y	0	Y	N
175.	Iran-Iraq	Oct 83	Y	1	N	N
176.	Korea-Burma	Oct 83	Y	1	Y	N
177.	Grenada	Oct 83	Y	1	Y	Y
178.	Syria	Dec 83	Y	1	Y	N
179.	El Salvador	Jan 84	N	0	Y	N
180.	Central America	Mar 84	Y	1	Y	Y
181.	Egypt/Libya/Sudan	Mar 84	Y	0	Y	N
182.	El Salvador	Mar 84	N	0	Y	N
183.	Persian Gulf	Apr 84	Y	1	Y	N
184.	Saudi Arabia	Jun 84	N	0	Y	N
185.	Red Sea Mines	Aug 84	Y	0	Y	N
186.	Sudan-Chad	Aug 84	N	0	Y	N
187.	Beirut Embassy	Sep 84	Y	0	N	N
188.	El Salvador	Oct 84	N	0	Y	N
189.	Colombian Embassy	Nov 84	N	0	Y	N
190.	Saudi Hijacking	Nov 84	Y	1	N	N
191.	Cuba	Nov 84	Y	1	Y	N
192.	U.S. Personnel in Lebanon	Mar 85	Y	1	N	N
193.	TWA 847 Hijacking	Jun 85	Y	1	N	N

Table B.1—continued

No.	Operation/Event/Location	Date Begun	USN Involved?	If So, No. of CVs Used	USAF Involved?	U.S. Army Involved?
194.	Persian Gulf	Sep 85	Y	0	N	N
195.	SS *Achille Lauro*	Oct 85	Y	1	N	N
196.	Egypt Air Hijacking	Nov 85	Y	1	N	N
197.	Persian Gulf Escort	Jan 86	Y	0	N	N
198.	Yemen Civil War	Jan 86	Y	0	N	N
199.	OVL-FON Ops	Feb 86	Y	3	N	N
200.	Lebanon Hostages	Mar 86	Y	0	N	N
201.	Libya Strike	Apr 86	Y	2	Y	N
202.	Pakistan Hijacking	Sep 86	Y	1	N	N
203.	Korea	Sep 86	N	0	Y	N
204.	Persian Gulf Ops	Jan 87	Y	2	Y	Y
205.	Hostages in Lebanon	Feb 87	Y	1	N	N
206.	Haiti	Jan 88	Y	0	N	N
207.	Honduras	Mar 88	N	0	Y	Y
208.	Panama	Apr 88	Y	0	Y	Y
209.	Pakistan	Apr 88	N	0	Y	N
210.	Summer Olympics	Sep 88	Y	2	Y	Y
211.	Burma Unrest	Sep 88	Y	0	N	N
212.	Maldives Coup	Nov 88	Y	1	N	N
213.	Lebanon Civil War	Feb 89	Y	1	N	Y
214.	Panama Elections	May 89	Y	1	Y	Y
215.	Pakistan-Afghan	May 89	N	0	Y	N
216.	China Civil Unrest	Jun 89	Y	1	N	N
217.	Hostages in Lebanon	Aug 89	Y	2	N	N
218.	Philippines	Nov 89	Y	2	Y	N
219.	Panama	Dec 89	Y	0	Y	Y
220.	Liberia NEO	May 90	Y	0	N	N
221.	Iraqi Pressure on Kuwait	Jul 90	Y	0	Y	N
222.	Operation Desert Shield	Aug 90	Y	6	Y	Y
223.	Trinidad Coup	Aug 90	Y	0	N	N
224.	Somalia Evacuation	Jan 91	Y	0	Y	N
225.	Israel	Jan 91	N	0	N	Y
226.	Sudan NEO	Jan 91	N	0	Y	N
227.	Postwar Iraq Sanctions	Feb 91	Y	1	N	N
228.	Provide Comfort	Apr 91	Y	1	Y	Y
229.	Haiti	Sep 91	Y	1	Y	Y
230.	Zaire	Sep 91	N	0	Y	N
231.	Sierra Leone	May 92	N	0	Y	N
232.	Iraq-Kuwait	Aug 92	Y	1	Y	Y
233.	Southern Watch	Aug 92	Y	1	Y	Y
234.	Combat, Search and Rescue (CSAR) in Bosnia—Italian Pilot	Sep 92	Y	0	N	N
235.	Liberia Evacuation	Oct 92	Y	1	Y	N
236.	Tajikistan NEO	Oct 92	N	0	Y	N
237.	Sharp Guard	Jul 92	Y	0	N	N
238.	Somalia—UN Support	Sep 92	Y	0	Y	N
239.	Restore Hope —Somalia	Nov 92	Y	1	Y	Y
240.	Iraq Strikes	Jan 93	Y	0	Y	N
241.	Haitian Maritime Intercept	Jan 93	Y	0	N	N
242.	Deny Flight	Apr 93	Y	1	Y	N

Table B.1—continued

No.	Operation/Event/Location	Date Begun	USN Involved?	If So, No. of CVs Used	USAF Involved?	U.S. Army Involved?
243.	Iraq Strikes	Jun 93	Y	1	Y	N
244.	Macedonia	Jul 93	N	0	N	Y
245.	Somalia—Withdrawal	Feb 94	Y	0	N	N
246.	Rwanda—U.S. Citizen Evacuation	Apr 94	Y	0	Y	N
247.	Haiti Embargo	May 94	Y	0	N	N
248.	Yemen NEO	May 94	Y	0	Y	N
249.	Iraq-Kuwait	Jun 94	Y	1	Y	Y
250.	Rwanda Intervention	Jun 94	N	0	Y	Y
251.	North Korea Tensions	Jun 94	Y	1	Y	Y
252.	Haitian Intervention	Sep 94	Y	2	Y	Y
253.	Bosnia Strikes	Sep 94	Y	1	Y	N
254.	Somalia—U.S. Liaison Office Evacuation	Sep 94	Y	0	Y	N
255.	Somalia—Final Evacuation	Jan 95	Y	0	Y	N
256.	Cuban Shootdown	Feb 95	Y	1	N	N
257.	Bosnia—NATO Support	May 95	Y	1	Y	N
258.	CSAR in Bosnia	Jun 95	Y	0	Y	N
259.	Bosnia Strikes	Aug 95	Y	1	Y	N
260.	Iraqi Defectors to Jordan	Aug 95	Y	1	N	N
261.	Bosnia—NATO Intervention	Dec 95	Y	1	Y	Y
262.	Bosnia—Headquarters Security	Dec 95	Y	0	N	N
263.	China-Taiwan	Feb 96	Y	2	N	N
264.	Liberia NEO	Apr 96	Y	0	Y	N
265.	Central African Republic	May 96	Y	0	Y	N
266.	Haiti—UN Security	Jul 96	Y	0	N	Y
267.	Iraq-Kurdish Conflict	Sep 96	Y	2	Y	N
268.	Burundi NEO	Sep 96	N	0	Y	N

CSAR = Combat Search and Rescue; NEO = Noncombatant Evacuation Order; Ops = Operations; PRC = People's Republic of China; ROC = Republic of China.

Appendix C

CARRIER FLEETS OF THE WORLD

The U.S carrier fleet dwarfs the fleets of other nations in size and number of ships, and number of aircraft embarked (see Table C.1). This does not mean that the benefits of fielding a major carrier force have been lost on the rest of the world. Other nations have sought to acquire and operate carriers in support of both military and political goals—a difficult and expensive venture. Britain's carrier fleet in particular has been drastically altered since the 1960s, mostly for budgetary reasons. Despite its reliance on submarines and cruise missiles, the Soviet Navy attempted, with only partial success, to create a large-deck carrier force as the only method of challenging American naval dominance. The only other nations currently operating conventional-take-off-and-landing (CTOL) vessels are Argentina, Brazil, and France. The first two operate primarily anti-submarine warfare (ASW) aircraft, with a small number of A-4 or Super Etendard fighter/attack jets. The other nations listed have vertical/short-take-off-and-landing (V/STOL) carriers (built as carriers or modified from other ships) that operate helicopters and Harrier attack planes.

Only the smaller V/STOL ships are of new construction, and how long the older vessels can be maintained in service is problematic. While the newer ships are quite capable in certain scenarios, they are generally rather limited. Although the Royal Navy's performance in the 1982 Falklands War was successful by most measures, it also highlighted the limitations of the V/STOL carrier and aircraft. The conflict was more protracted than expected because the British lacked credible airborne early-warning and electronic warfare capabilities, and sufficient long-range fighter and strike aircraft.

France remains the only other nation with the wherewithal to field modern seaborne aviation. It has a nuclear-carrier construction program and new aircraft in development. Still, size and capabilities will be closer to those of the old Essex class or, at best, Midway class than to those of current U.S. vessels.

Table C.1

World Carrier Fleets, 1996

Nation	Class	Displacement (tons)	Number of Aircraft
Argentina	25 de Mayo[a]	20,000	21
Brazil	Minas Gerais	20,000	20
France	Clemenceau (2)	32,000	35–40[b]
	Charles de Gaulle[c]	40,000	35–40
Great Britain	Invincible (3)[d]	20,000	14
India[e]	Vikrant	20,000	16
	Viraat	28,000	20
Italy[f]	Giuseppe Garibaldi	14,000	16–18
Russia	Kiev (3)	43,000	33[g]
	Admiral Kuznetsov (2)[h]	59,000	40–60
Spain	Principe de Asturias	17,000	17
Thailand	Chakri Naruebet	11,500	12–15
United States	Independence	81,000	75[i]
	Kitty Hawk (3)	81,000	
	Enterprise	94,000	
	Nimitz (7)[j]	91,000–102,000	

[a]Not operational since 1985. Primary reason for continued existence is to justify a fixed-wing naval-aviation component.

[b]Acquisition of E-2 airborne early-warning aircraft and development of Rafale M naval strike fighter will give France the most capable sea-based air force outside the United States.

[c]Lead ship of projected class of two nuclear-powered vessels. Construction has slid 3 years for budgetary reasons; sea trials now projected for 1999, at the earliest.

[d]Only two of the three are operational at any one time; the third is in preservation status until another requires extended maintenance.

[e]Plans to build two new carriers have not materialized, for economic reasons. It is possible New Delhi may purchase a French Clemenceau or one of the Russian Kievs.

[f]Italy will undertake construction of a new CVS of V/STOL design.

[g]The V/STOL Forger naval strike-fighter was removed from service in 1992. This class is primarily a helicopter carrier.

[h]One additional ship under construction will likely not be completed.

[i]Nominal air wing consists of 50 strike and fighter aircraft, with approximately 25 for early warning, electronic warfare, ASW, tanking, synthetic aperture radar, and logistics.

[j]Two more Nimitz-class vessels are currently under construction.

COMPONENTS OF SHIPBUILDER COST DIFFERENCES

In Chapter Four, we show how different start dates and build periods affect total shipbuilder cost for CVN 77. Here, we break down these differences into shipyard labor (including variable overhead), fixed shipyard overhead, and contractor-furnished equipment (CFE) costs (including the costs of raw materials and purchased parts not typically considered CFE). For convenience, we first repeat Figure 4.10, which shows differences in total cost, as Figure D.1. Figures D.2, D.3, and D.4 illustrate the contributions of the three sources of cost to those differences in total cost (all graphs are to the same scale).

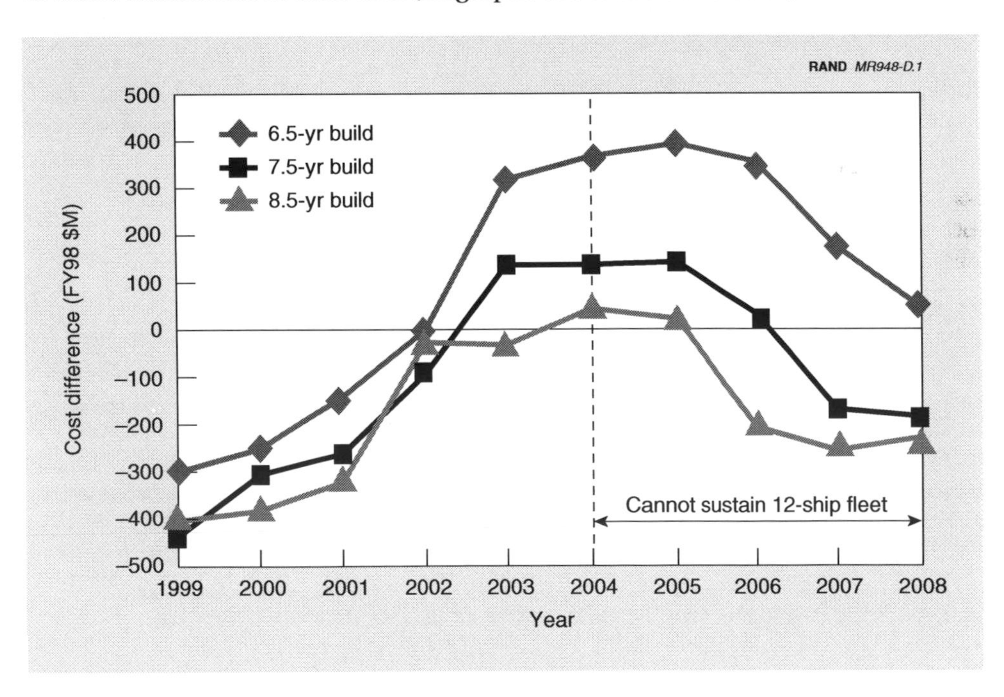

Figure D.1—Effect of CVN 77 Start Date and Build Period on Total Shipyard Costs

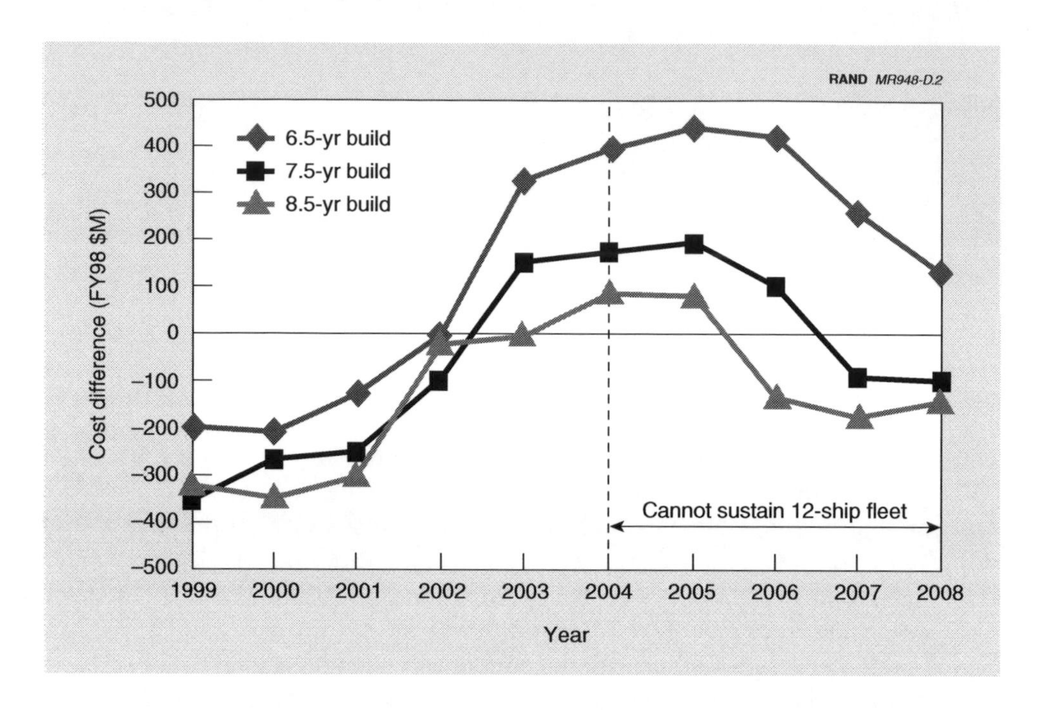

Figure D.2—Effect of CVN 77 Start Date and Build Period on Shipyard Labor Costs

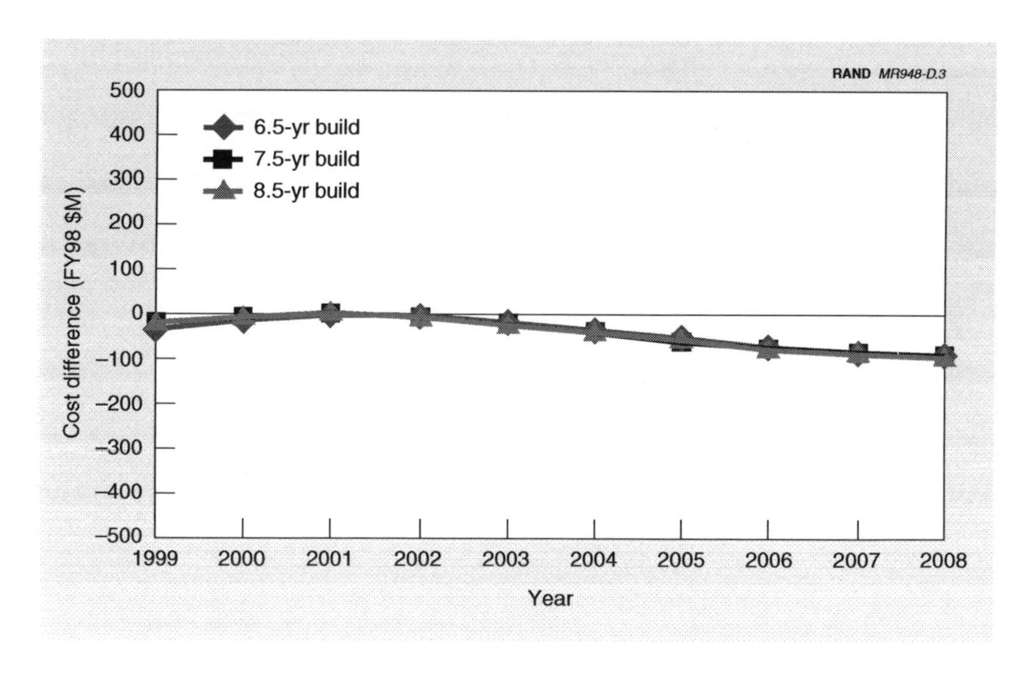

Figure D.3—Effect of CVN 77 Start Date and Build Period on Shipyard Fixed-Overhead Costs

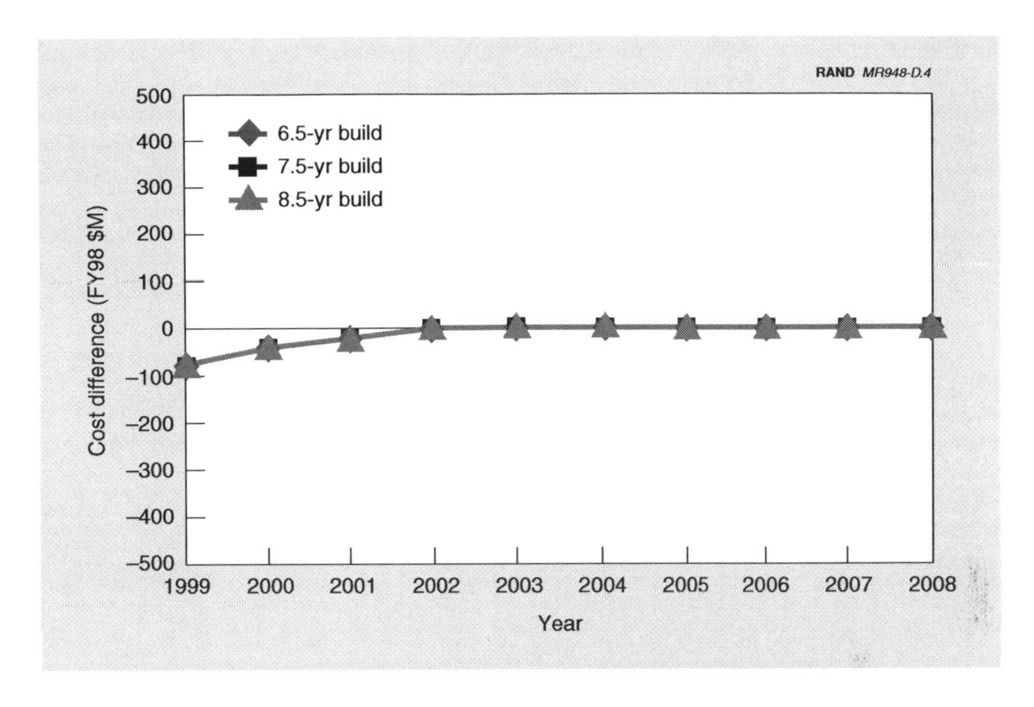

Figure D.4—Effect of CVN 77 Start Date and Build Period on Shipyard Costs for Contractor-Furnished Equipment

There is not much difference between the total-cost and labor-cost curves, because the fixed-overhead and CFE profiles do not vary much from the baseline and do so at different times. Figure D.3 shows the effect of different dates and build periods on the portion of NNS's fixed-overhead costs attributed to CVN 77. The shipyard's fixed overhead includes depreciation, amortization, taxes, etc., and therefore does not change with yard workload. The portion of fixed overhead attributed to CVN 77 is affected by the amount of other work that is in the yard at the same time; if there is more work, CVN 77 bears a smaller share of fixed overhead. Thus, the later the start date beyond 2001, the greater is the average amount of work in the yard over the build period and the smaller is the portion of the yard's fixed overhead that CVN 77 must bear.

It should be kept in mind that the Navy generates the vast majority of the work at NNS. Thus, the Navy does not necessarily save overhead expense by postponing CVN 77. Whatever the CVN 77 program does not pay for will most likely be covered by another Navy program.

Figure D.4 shows the effect of different dates and build periods on the costs of construction materials and intermediate products provided to the shipyard by vendors. These costs are included in the contract let by the government to the shipyard. They do not include the costs of government-furnished equipment, such as the reactor. The CFE cost differences over time are based on a survey of vendors conducted by NNS in fall 1996.

Appendix E

LABOR REQUIRED FOR VARIOUS PROJECTS AT NEWPORT NEWS SHIPBUILDING

In Chapter Four, we compared the labor-demand profile for a product tanker with that expected for CVN 77 to make the point that commercial work could not substitute for a carrier in leveling the shipyard's workforce. In this appendix, we present a full range of labor-demand profiles for various projects, all to the same scale. The legend for all is the same as that for Figure 4.1. Abbreviations are as follows:

SRA	Selected restricted availability
ESRA	Extended SRA
EDSRA	Extended docking SRA
DPIA	Docking planned incremental availability
RCOH	Refueling/complex overhaul
PSA	Postshakedown availability
NSSN	New attack submarine (type 1 is first few ships of class; type 2 is remainder).

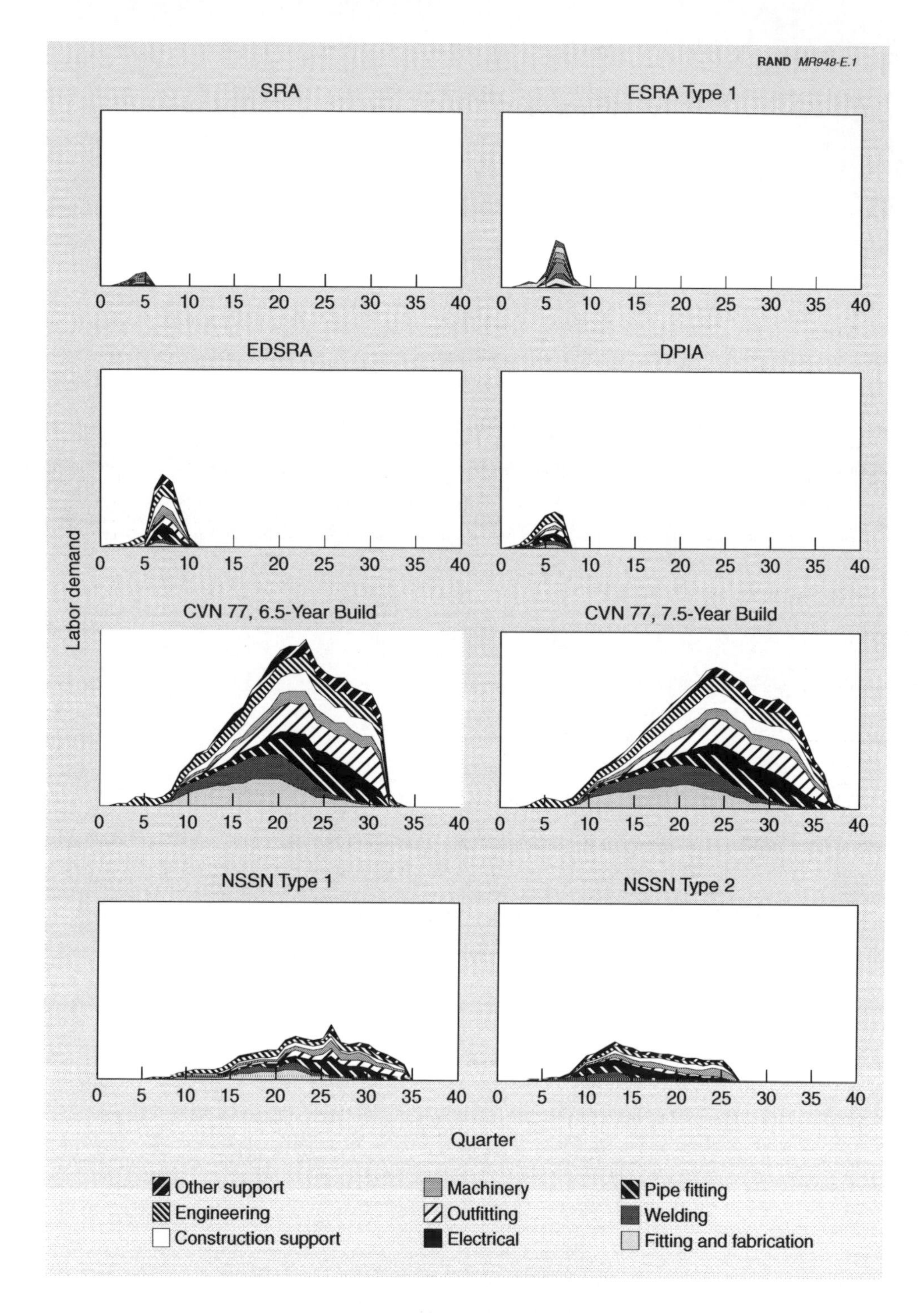

Figure E.1—Labor-Demand Profiles for Various Projects in Newport News Shipyard

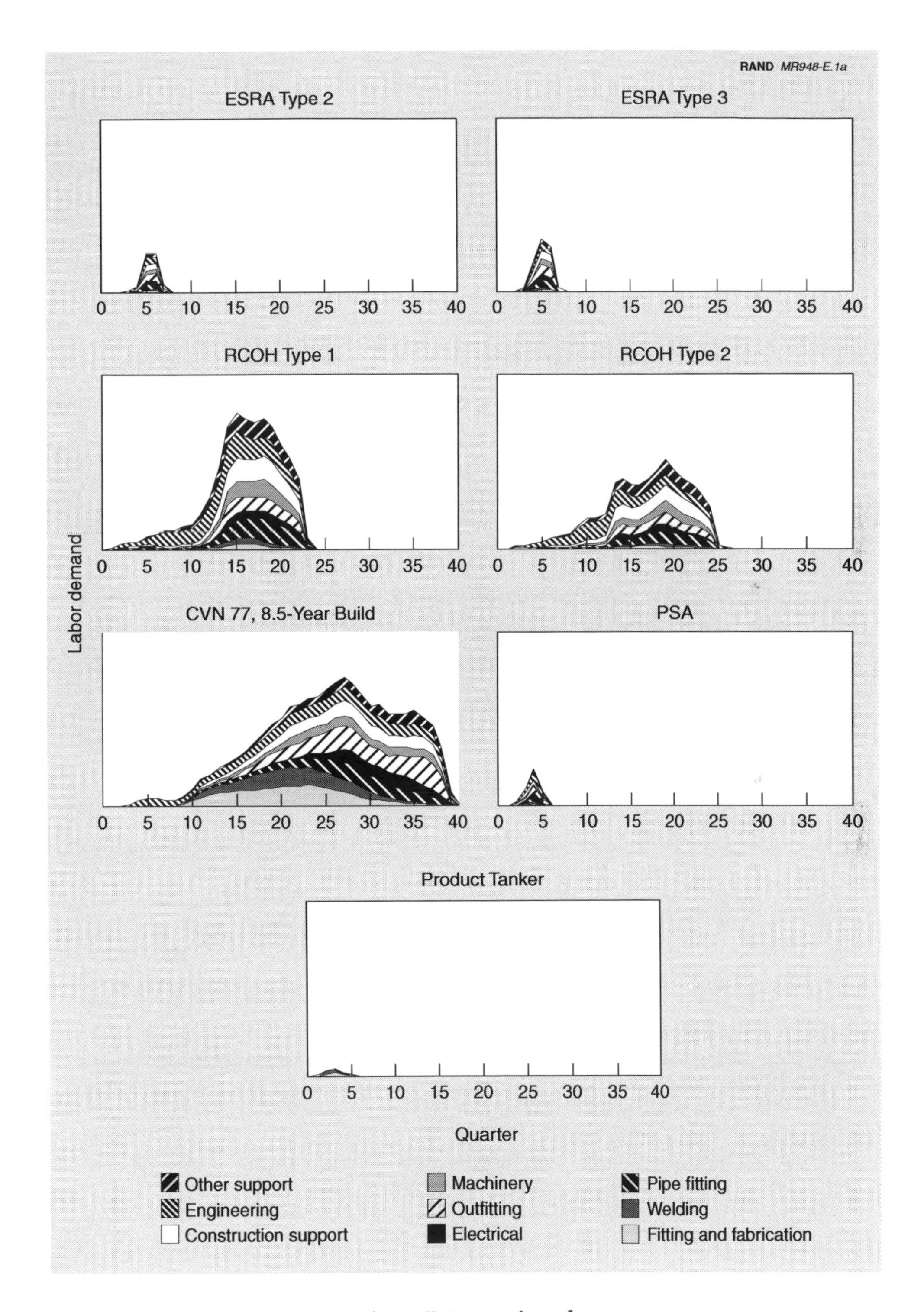

Figure E.1—continued

LARGE SURFACE-SHIP PRODUCTION IN GREAT BRITAIN

This appendix documents the findings from a series of meetings with the Ministry of Defense (MoD) of the United Kingdom (UK) and with representatives of shipyards in that country: Vickers Engineering and Shipbuilding Limited (VSEL) and Kvaerner Govan Limited (KGL).

STRUCTURE OF THE BRITISH AIRCRAFT CARRIER FORCE

The Royal Navy currently has three Invincible-class (CVS) aircraft carriers in its force structure. These carriers have an approximately 20,600-ton displacement fully loaded and typically carry a complement of seven Sea Harrier FA-2 short take-off/vertical landing (STOVL) aircraft, plus seven anti-submarine warfare (ASW) Sea King HAS-6 and three airborne early warning (AEW)-2 Sea King helicopters.[1] The lead ship in the class was designed and built by VSEL at Barrow-in-Furness, England; the remaining two were built to the VSEL design and blueprints by Swan Hunter Shipbuilders at Wallsend, England. The timelines for these carriers are shown in Table F.1.

Table F.1

Invincible-Class Timelines

Ship Name	Number	Builder	Date Keel Laid	Date Launched	Date Commissioned
Invincible	R 05	Vickers	20 Jul 1973	3 May 1977	11 Jul 1980
Illustrious	R 06	Swan Hunter	7 Oct 1976	1 Dec 1978	20 Jun 1982
Ark Royal	R 07	Swan Hunter	14 Dec 1978	2 Jun 1981	1 Nov 1985

SOURCE: *Jane's Fighting Ships*, London: Sampson Low, Marston and Co., 1996.

[1]Under surge conditions, the Invincible class can embark 12 Harriers and seven ASW and three AEW Sea King helicopters. Under either normal or surge conditions, approximately 15 of the aircraft and helicopters are positioned on the hangar deck of the ship.

Only two of the three carriers are in operational status at any one time; the third is in a semi-mothball status or in overhaul. This arrangement is due primarily to budget, personnel, and aircraft constraints. For each ship, there is an approximate 10-year cycle between modernization and overhaul.

The British MoD is currently in the initial stages of planning for the next class of carriers; the anticipated delivery is approximately 2012. Therefore, there will be approximately a 20-year gap in the production of aircraft carriers in Great Britain. However, two major issues are currently unfolding in Great Britain that will affect the design and construction of large military ships:

1. The carriers are in the 20,000-ton range, and two other classes of ships (amphibious ships) of roughly the same size have been

   - designed (landing platform–helicopter [LPH] and landing platform–dock [LPD])
   - are under construction (LPH—design start in 1993, to be commissioned in 1998)
   - are about to start construction (LPD).

2. The military shipbuilding environment in Great Britain has undergone major changes so that

   - the traditional MoD carrier knowledge base is quickly becoming less important
   - the lack of continuity in shipyard practices is even less important because commercial practices are inherently quite different.

Therefore, the gap in large ships is not as great as might be inferred from looking at carriers only.

In the following two sections, we discuss these issues in turn. We then detail specific implications these issues have for the two major shipbuilders, VSEL and KGL. We conclude with remarks on carrier-specific considerations.

LPH AND LPD SHIP CLASSES

With regard to the first issue that affects design and construction of military ships—size—the British Royal Navy is currently in the process of acquiring LPH and LPD ships of approximately the same size as the CVS class.

Amphibious Helicopter Carrier (LPH)

The amphibious helicopter carrier, designated as a landing platform–helicopter, was originally conceived in the mid-1980s as a replacement for HMS *Hermes,* to provide the type of amphibious assault required during the Falklands campaign. The primary role of the ship is to carry an embarked military force of up to 800 personnel, 12 medium-support helicopters (currently Sea Kings, but EH-101s in the future), and 6 Lynx helicopters of the UK/Netherlands amphibious force. This force would then be disembarked as part of an amphibious assault. Secondary roles include afloat flight training, limited ASW operations with suitable helicopters, ferrying Sea Harriers, and being a base for anti-terrorist operations. Up to 20 Sea Harriers can be carried, but not supported.

The first competition for the design and build of the LPH (then called the auxiliary support ship [ASS]) was held in 1988. The Invitation to Tender (ITT; similar to a Request for Proposal [RFP] in the United States) included a price guide. Unfortunately, the shipyards bidding to construct the ship either offered a compliant design that was unaffordable or a design with major shortcomings that was affordable.

The MoD then conducted an Options for Change study (of redesign to meet cost constraints). In spring 1991, cost/capability studies indicated that a worthwhile ship was feasible at an affordable cost. The requirements were revised and a new ITT prepared, the ship's place in the defense program having been confirmed. The ITT was issued in February 1992, and tenders from VSEL and Swan Hunter Shipbuilders were received the following October. The tender assessment was accelerated in early spring 1993, when the opportunity for a prompt contract award became apparent. VSEL was awarded the design and build contract on May 11, 1993, having made a significantly cheaper best and final offer.

Swan Hunter Shipbuilders protested the award. Their protest led to a National Audit Office (NAO) inquiry into the conduct of the competition. The NAO concluded that the competition had been conducted competently and in a fair manner, and that, given VSEL's price advantage, the decision to award them the contract was correct. The NAO did make some recommendations for the management of the contract, emphasizing the importance of risk management and the rigorous control of change.

The design and build contract itself is different from contracts in past practice. It is a fixed-price contract and, since VSEL had almost exclusively built submarines (both diesel and nuclear attack, and nuclear ballistic missile submarines) for an extended time, VSEL subcontracted the construction of the basic hull to the Kvaerner Govan Limited (KGL) shipyard in Glasgow, Scotland. The Kvaerner Govan yard had experience in building large commercial ships but had never constructed a military ship of the size of HMS *Ocean*. The vessel steamed under its own power in November 1996, from KGL on the Clyde to VSEL in Barrow, where the ship is being outfitted with her military features and combat systems.

Constrained defense budgets caused the MoD to decide to use commercial standards in place of military specifications for the hull construction. However, commercial and military standards and equipment were very carefully considered and combined as needed to maintain naval functionality while controlling construction costs. As a result, the hull meets shock performance requirements. Elsewhere within the ship, systems have been designed to meet levels of military performance. For example, the firefighting system meets the full functionality of a naval system but consists of a mix of commercial and military equipment.

VSEL personnel were positioned at the Govan yard, and the VSEL Barrow facility fabricated some sections of the hull and transported them to the Kvaerner Govan yard. Through this experience, VSEL hoped to regain expertise in the construction of large surface ships.

The hull form of the LPH is based on that of the CVSs and has similar dimensions and displacement. Her main propulsion plant will be medium-speed diesels driving through two reduction gear boxes with fixed-pitch propellers.

The keel for *Ocean* was laid on May 30, 1994, and the ship was launched on October 11, 1995. Plans call for acceptance from VSEL following her Part IV Trials program (initial Royal Navy trials with crew) in spring 1998, and she will enter operational service in early 1999.

Landing Platform–Dock (LPD)

Under the Landing Platform–Dock Replacement (LPD-R) program, the MoD has recently placed a contract to construct two new LPD-class ships, which will replace the Royal Navy's HMS *Intrepid* and HMS *Fearless*. VSEL will design and build these ships at its Barrow facility. The company was the only remaining contender for the contract after a protracted design and competition phase started in 1991, following lengthy studies during the 1980s.

Displacing 13,000 tons, the new LPDs will have a speed of 18 knots, and each will carry 650 troops and 325 crew. They will have a flight deck that can operate two Merlin (EH-101) or Sea King helicopters or a single Chinook, and they will carry eight landing craft, four of which will be capable of landing main battle tanks. The ships, named HMS *Albion* and *Bulwark*, are expected to enter service after 2000, by which time the current LPDs will be almost 40 years old.

Like *Ocean*, the LPDs will incorporate commercial standards instead of military standards in their hull construction. VSEL designers worked very closely with the MoD to reduce the procurement cost of these new LPDs by approximately one-third of initial cost estimates. The two teams, VSEL and MoD, did so by going through the technical specifications line by line to reduce both risk and cost.

CHANGES IN MILITARY SHIPBUILDING IN GREAT BRITAIN

As with construction of *Ocean* and the LPDs, adoption of commercial practices is one of several major changes that figure in the current military shipbuilding environment in Great Britain. Driven by reduced defense budgets, the MoD military ship acquisition policy is now

- emphasizing competition
- shifting design and production risk from the government to prime contractors and subcontractors
- encouraging adoption of commercial practices, when practical
- allowing contractors more flexibility.

In response, UK defense industries are restructuring and combining their business units.

The MoD Acquisition Policy

Emphasizing Competition. Using competition to drive down procurement costs is a major theme throughout the MoD, and military shipbuilding is no exception. With the MoD's strong push for competition, the move toward commercial practices, the desire to have a prime contractor responsible for the entire ship design and construction, and the introduction of two large electronics-oriented organizations into military-ship construction, the traditional business model of military-ship acquisition is changing very rapidly.

As a result, partnerships of organizations, including some organizations with no shipbuilding experience, are now competing for new-ship construction.[2] For example, GEC, a large electronics, power systems, and telecommunications

---

[2]These partnerships are also being used in the United States, specifically for the LPD-17 program.

conglomerate,[3] teamed with a manufacturer of offshore oil rigs and a group of former MoD submarine engineers to bid against VSEL and other competitors on the Batch 2 Trafalgar submarine program. British Aerospace (BA) has also entered into competitions for new shipbuilding contracts. Owing to limited budgets, the MoD has strongly advocated competition and typically awards new contracts to the lowest bidder, conditional on the winning bidder's having sound plans for the design and construction program.

While the MoD would be willing to extend the logic of competition and use of commercial practices to have a foreign yard build the hull, politics would probably prohibit such a practice. Yet, in the LPH contract, discussed above, MoD did entertain a bid from Norway for a brief time.

Shifting Design and Production Risk to Prime and Subcontractors. The MoD is also shifting program risk to the prime contractor in any new-construction program. Many of the design and engineering functions are migrating from the MoD to the prime contractors. The use of modern computer software, including computer-assisted design (CAD) and virtual-reality systems, is helping both the MoD and the contractor examine various design alternatives before any actual production takes place. The assumption is that these advanced modeling skills will allow the MoD and the contractors to "do it right" during construction of the lead ship instead of having the lead ship serve as something of a learning experience for follow-on ships in the class.

One drawback of this new approach is the lack of adequate tools to estimate cost and schedule implications of various design options. The use of commercial products and standards is a new practice, and sufficient data do not exist to construct adequate cost and schedule estimates. Further, since the shipyards will have actual construction cost data but are not contractually obligated to provide those data to the MoD, the MoD may never have sufficient data to build good cost-estimating models. In many ways, the MoD is relying on experience in the commercial marketplace and on faith that "commercial is good enough" for many functions on board a military ship.

These changes are evident in the contract award and construction of the new LPH, HMS *Ocean*. Previously, the MoD would have performed analyses of the requirements, initial concept, and feasibility and collaborate on the project-definition phase with the shipyard. For *Ocean*, VSEL participated heavily in the concept and feasibility analysis and performed the majority of the project-definition function. In fact, VSEL was fully responsible for the complete design

---

[3]GEC has acquired VSEL and has formed a GEC Marine Division composed of VSEL, Yarrow Shipbuilders Limited, and the National Nuclear Corporation. Therefore, VSEL is now part of GEC, a competitor in the original bidding for the Batch 2 Trafalgar (nuclear-submarine) program.

of the LPH from concept to detailed design. Also, for prior ships, the MoD would have performed the test and acceptance function once the ship was constructed. For *Ocean*, VSEL will perform that function. In this current shipbuilding environment, the MoD does very little design work, allowing the prime contractor or shipyard to perform the design function.

Also very different from past practices is the contract for *Ocean*, which is fixed price. The contractor is encouraged to take full advantage of the cost reductions associated with commercial standards and practices as long as those commercial standards do not jeopardize the military operational missions or the safety of military personnel.

Since VSEL had been building exclusively submarines and, hence, has been out of the surface-ship business for many years (see the "Implications for Shipbuilders" section on VSEL) and has had little current commercial experience, it decided to subcontract the hull building (steel erection) to Kvaerner Govan in nearby Glasgow. The MoD is working closely with VSEL and, through them, with KGL, on the hull build.

MoD officials indicated that the process is still very much in flux, but that they are trying to make good business decisions, banking on the rationale that if something is good enough to be commercial grade, it cannot be overlooked. Of course, when commercial is not sufficient to meet their needs, they permit a "MILSPEC" item or process.

Since the ship will be "built in a computer" (CAD/CAM), and since the building yard will own the computer model, the building yard will be the logistics and maintenance yard for the life of the ship—unless the MoD pays for the model and its transfer to another facility, which is unlikely—so the competition to build may turn out to be (de facto) a competition for cradle-to-grave support.

While this approach is very conducive to cost as an independent variable (CAIV) application, other conditions may force a return to more-traditional practices. There were no competitive bids for the LPD. The only bid (VSEL) was too high for the MoD. To reduce the total cost, the MoD and the yard, together, analyzed each cost item to see what could be done. The contractor had provided a cost that included risk money, because it had been asked to translate a short, "functional" specification (for example, "be able to fight 5 fires at once") to a "build" specification ("so many pumps," "this kind of pipe," etc.). MoD opted to move away from a functional specification and agreed with the builder on a design that would prove adequate. Consequently, contractor risk is reduced and the price drops. By this method, the MoD took out 33 percent of the cost, but, in some areas, had to become involved with the detailed (build) specifications rather than staying with the functional or performance specifications.

Encouraging Adoption of Commercial Practices. The MoD is aware of the risks involved with this new way of doing business and is being careful to make sound military decisions within the fiscal constraints. The move to commercial standards and practices (which has an analog in U.S. acquisition reform initiatives, which encourage "commercial-off-the-shelf" [COTS] products when appropriate) is being monitored closely. The basic assumption is that commercial products must work to remain competitive in the civilian marketplace. However, usually no data have been available for measuring the reliability of the commercial systems, especially in a military environment, and there is always the chance that commercial suppliers may go out of business, potentially creating logistics problems for the military users (this is also a potential problem with suppliers of military products).

The MoD estimates that the new computer design technologies (CAD/CAM), combined with a strong effort to simplify the ship and with "going commercial," results in about a 33-percent reduction in man-hours to build, because the ship is built right the first time. It also believes that while the shipbuilders think they are being efficient, or are becoming more so, there is still more efficiency to be gained. An internal VSEL study suggested that personnel were actually working less than half the time, and waiting the rest. But the core capability of shipyard personnel is integration; and a time-and-motion analysis of people in the yard may not capture the essence of the job.

The MoD is aware that commercial suppliers may go out of business and not be there for the long-term logistics. They hope that other options will arise to help them fix problems if and when such problems arise.

Allowing Contractors More Flexibility. MoD pointed out that, contrary to the Naval Sea Systems Command (NAVSEA) practice of "affordability through commonality" (i.e., economies of scale), the number of ships in the Royal Navy is not sufficient to justify such an approach. As well, because MoD is using a greatly diverse supplier base to foster competition, although ships of the same class would be unlikely to have different equipment, different suppliers could provide similar equipment across ship classes (for example, laundry or compressors). And although the MoD does give the prime contractor flexibility in selecting equipment, builders of follow-on ships in a class must bear the full cost of introducing any new equipment into the design and construction of the ship.

MoD has paid special attention to simplifying the ship design so that "going commercial" provides the biggest benefit.

These two issues have had implications for both major shipbuilders. We discuss those implications for VSEL and Kvaerner Govan in the next two sections.

IMPLICATIONS FOR SHIPBUILDERS—VSEL's EXPERIENCE

Prior to being nationalized in the 1980s, VSEL employed 14,000 to 15,000 people in producing surface ships, submarines (nuclear-powered and conventional), and other major items (for example, guns and locomotives). When it was nationalized by the government, VSEL was directed to divest its surface-ship capabilities and concentrate on building submarines. To implement this objective, they built a land-level facility (similar to those at Electric Boat and Newport News Shipbuilding), optimized their production methods for submarines, started modular construction, and ensured that their land-level facility was set up in an efficient way to reduce cost.

A few years later, they were privatized and were asked to compete in the broader shipbuilding world. "Going commercial" was a key part of the competition. VSEL is undergoing major changes in trying to adapt to the new world. Now employing only about 5,300 people, VSEL finds that the overhead of the land-level facility and of maintaining a nuclear license makes it difficult for it to be competitive in the current environment. VSEL is working with the MoD to allocate that overhead to a special account so that it can compete for other contracts. The overhead for the nuclear-license part may be considered by the MoD, but the huge land-level-facility part is still under discussion.

Subcontracting to Redevelop Full Capability

To redevelop full shipbuilding capability, VSEL has expanded into the surface-ship world—a reason it bid on the LPH contract and subcontracted with Kvaerner Govan for the hull. VSEL is carefully monitoring that process to learn how to rebuild surface ships and how to be efficient in a commercial environment. It has subcontracted some simple steel-erection work from KGL to its own Barrow yard so that it can apply the lessons learned. While this learning process is going on, VSEL considers its efforts (and losses or meager profits) an investment in the future of the company. On the subcontract to KGL, it is not making much (if any) profit but has learned enough to feel that it can now be competitive in surface-ship commercial practices.

Over the years, the yard layout has been optimized for submarines. With the new surface-ship contracts, getting a ship of the planned size through the yard and to the water requires tearing down parts of buildings in town and working with the city to widen public streets to move large sections of the hull. Modifying the yard was not economical for a single ship, but, with the winning of the contract for the two LPDs, the modifications to the yard and public streets will be made.

VSEL's primary strength is in performing ship integration. For more-complex surface ships and for submarines, VSEL believes that it has unique skills that will help it succeed in the long run. However, to get to the "long run," it needs to survive by doing all jobs (even at little or no profit) that contribute to its long-term strategy of becoming a full-capability yard that can handle both military and commercial work. VSEL believes that it is accomplishing its goal: Despite subcontracting with Kvaerner for the steel erection of the LPH, it considers itself ready to build the LPD on its own.

Competition Versus Core Capability

Emphasizing on several occasions that competition is fine and that it is willing to compete, but realizing that a company must really know the product in order for competition to be fruitful for the MoD, VSEL clarified its knowledge by discussing the LPD design (provided by a design firm) it was asked to build. There were several problems, most significant being the lack of design margin necessary to account for changes and growth during the life of the ship. The design team obviously lacked experience and sufficient knowledge to question the MoD's intentions. As a result, VSEL had to make several modifications (with MoD's participation) before it could provide a cost bid. The implication was that competition for its own sake and a bid by a new entrant in the field may be interesting, but that, in the long run, you have to pay the real cost of what you are buying.

We discussed the bidding war on Arleigh Burke–class destroyers (DDGs). VSEL managers indicated that they were concerned that too much emphasis on competition was driving them in that direction, but that MoD may not fully understand all the adverse implications.

VSEL has embraced the concept of competition. While learning how to work in a more commercial environment and willing to optimize costs by subcontracting extensively in most areas, it wants to maintain integration and outfitting as a core capability. It is burdened by the overhead items from its nationalized days but is working on fixes with the MoD. Facilities modifications to accommodate the new type of work are going to be expensive.

Skilled personnel flow to and from VSEL much as engineers flow, with contracts, into and out of aerospace companies in the United States. Therefore, maintaining specific skills is not as critical as it once was. And geographical separation of the various yards is not as great in the United Kingdom as it is in the United States.

VSEL was not perceived to be big enough to be a prime contractor to the MoD. But with its acquisition by GEC, it is again viewed as a major competitor for new-ship construction. VSEL still considers the yard to have the key ingredient—integration skill—but welcomes the combat-system expertise that GEC brings.

It appears that taking the direction to specialize in submarines (during the nationalized period) was quite damaging and that going back to a broad-based-capabilities yard is both difficult and mandatory. VSEL has identified no "silver bullets" to aid in reaching its goal. It appears that hard-nosed business decisions are the operating principle. VSEL's relationships with its owners (GEC) and customers (the MoD at the moment) will be unfolding—reaching agreement on core competencies and realizing the benefits of competition.

IMPLICATIONS FOR SHIPBUILDERS—KVAERNER GOVAN's EXPERIENCE

The Kvaerner Govan Limited shipyard is situated on the south bank of the River Clyde in Glasgow, Scotland. The yard itself has been in existence for well over a century and was acquired in 1988 by Kvaerner Govan, a large, diverse Norwegian company with interests in energy, pulp and paper, engineering, and shipbuilding.

Considered the largest merchant shipbuilder in the UK, KGL has about 1,500 employees and broad experience in designing and constructing a large variety of vessels, such as bulk carriers, product tankers, container ships, and cruise ferries. In recent years, KGL has built only merchant ships and has specialized in the "high end" of shipbuilding: relatively complex ships such as chemical tankers and liquid-gas carriers. It is currently building a missile-transport tracking-and-command ship for a sea-launched missile program that can place commercial satellites in orbit

KGL constructed the hull and installed the propulsion and auxiliary propulsion equipment on the LPH, under a subcontract from VSEL. It made maximum use of commercial standards and practices to save construction costs. Once the hull and machinery were completed, the ship transited under its own power from Glasgow to the VSEL yard at Barrow-in-Furness for final outfitting with military features, installation and test of the combat system, and final trials. The completion letter was signed on November 25, 1996, and the ship is now at VSEL.

Overview

As a corporate policy, KGL would never enter a warship contract itself; however, it did undertake the subcontract from VSEL. KGL was concerned that a standard government contract would inundate and burden it with paper, inspection requirements, documentation, and the like. The negotiations with VSEL produced a one-volume specification, instead of the multiple volumes MoD imposed upon VSEL. All contact contractually was to be between KGL and VSEL, with no official contact between MoD and KGL. MoD and VSEL had a team of some eight overseers on-site throughout the construction period, acting as Supervisor of Shipbuilding/owner's representative.

The conceptual design of the ship was performed by VSEL and was provided as a package to KGL as part of the contract proposal. KGL would accept no classified drawings, and none was required, except for the hull lines and general arrangements. These drawings were kept in a safe and used only for reference. The detailed design was performed by KGL.

The hull and machinery were constructed and installed to the extent practical in accordance with rules for commercial shipbuilding laid down by Lloyd's Register of Shipping, the UK's classification-society counterpart of the American Bureau of Shipping (ABS). The configuration of the LPH, with large cut-out sections in its hull for aircraft elevators and landing craft, required a modification of the Lloyd's classification and certification process. The structure conformed to Lloyd's rules up to the tank tops; above the tank tops, Lloyd's Register accepted the results of finite-element analyses, as long as the local stresses and stress distribution met acceptable standards. A formal certification was not issued; instead, Lloyd's provided a letter stating that the ship, as modified, met its rules.

Wherever practical, commercial equipment and practices were utilized. The majority of construction cost savings can be attributed to "commercialization."

Subcontracting

Seventy percent of the hull, mechanical, and electrical (HM&E) work was done by KGL; the remaining 30 percent was contracted out. Turnkey operations were called for in the following areas:

- Hull thermal insulation
- Heating, ventilation, and air conditioning (HVAC)
- Firefighting systems
- Remote operating valves.

Electrical-power and distribution systems were contracted out but were not turnkey. Similarly, the laundry was not turnkey. KGL agreed that more turnkey operations were desirable and, indeed, was moving in that direction.

Commercial Instead of Military Specifications and Standards

Lloyd's Register of Shipping (LR) rules, which govern commercial-ship hull and machinery, were imposed, but MoD modified them to suit military requirements. Some of the rules and modifications are as follows:

- LR requires navigational lights to have dimmers; MoD did not want dimmers.
- LR allows interrupted welding of stiffeners; MoD required continuous welding.
- LR allowed thicker plates but fewer stiffeners. This was accepted by MoD and resulted in less deformation of the structure ("hungry horse" look) and a flatter flight deck.

MoD required an International Maritime Organization (IMO) "certificate" for the sewage system, but IMO has established no such certification. MoD requirements for piping were not as stringent as those the tanker buyers and operators imposed on KGL for chemical tankers.

Military specifications for electrical cabling were beyond commercial practices for redundancy and watertight integrity. As a result, KGL underestimated the man-hours required to pull the cables. MoD was not in favor of commercial standards for firefighting systems and made several changes during construction.

Commercial ships employ far more automation than that specified by MoD. KGL believes it could actually have installed more to commercial standards for the same price, or even cheaper. MoD did allow limited use of commercial pipe and fittings in the sewage system; more cost savings could have resulted from wider usage. MoD specified the use of sheet metal (light-gauge steel) for berths and lockers instead of commercially utilized wooden laminate.

Material and Equipment Procurement

KGL was able to purchase material and equipment far cheaper than MoD could, even for the very same item. Military specifications caused suppliers to charge twice as much as for making the same item to commercial specifications.

VSEL selected 1970s' model Pielstick engines, probably because those engines were already installed in other Royal Navy ships and, therefore, offered an overall lower cost than did other engine alternatives. KGL stated that it could have purchased a current model of Wartsilla engines at a much lower cost.

Manual, remotely operated valves, specified by MoD, are not in common commercial usage; they have been replaced by electro-mechanical control valves. Consequently, only one manufacturer could be found to supply the valves, and his price had to be accepted.

Life-Cycle Costs

A current-model Wartsilla diesel engine, which is cheaper to purchase than the specified Pielstick, is also 20 percent more fuel-efficient and would result in a direct 20-percent savings in fuel costs over the life of the ship. Additionally, the Wartsilla engines are supported by parts and service worldwide. The older engines would be more difficult and more expensive to support.

Paint systems[4] for tanks are available commercially and, if applied to ballast tanks, would last for the life of the ship. Although KGL has installed such paint systems on the chemical tankers it has recently constructed, it was apparent to KGL that the effects of corrosion were far more important to its commercial customers than they were to MoD.

More-extensive use of automation would require less crew. An integrated bridge with full control of the ship is a good example.

The heating, ventilation, and air-conditioning system for HMS *Ocean* was designed to maintain environmental conditions for electronics and computing equipment, but applied to the whole ship. KGL felt that a more balanced design would save not only operating costs but acquisition costs as well.[5]

Lessons Learned

KGL learned several lessons as it constructed HMS *Ocean* and reported them during our discussion:

1. Much tighter change control is necessary to minimize disruption and unreimbursed costs. This is difficult at times, especially when there are three parties involved—MoD, VSEL, and KGL—with communication often being two-way instead of three-way. Many of the changes were initiated by VSEL, not necessarily by the MoD. The MoD's change control on the prime contract with VSEL was, for the most part, very tight.

---

[4]A *paint system* is more than just a can of paint. Applications are designed and built on what is being painted.

[5]MoD may be moving toward greater recognition of life-cycle costs. New auxiliary-oiler contracts have higher paint-life requirements and specify an integrated bridge.

2. It is difficult to work to two standards in one shipyard. Even for HMS *Ocean,* with its large portion of commercial standards and specifications, KGL found that it could *not* readily and easily exchange workers from HMS *Ocean* to commercial projects. This is also a problem for U.S. yards engaged in naval and commercial ship construction.

3. KGL is headed toward using more and more subcontractors who will be performing operations with well-defined boundaries in a turnkey fashion.

4. KGL subcontracted the design of the electrical-power generation and distribution system to Siemens and the installation to a Norwegian company. In retrospect, KGL stated that the installation would have been far more efficient if Siemens had done the entire job. In the future, KGL intends to have electrical work done entirely by one subcontractor on all complex projects.

5. The preponderance of cost savings by constructing to essentially commercial standards in a commercial yard came from steel fabrication and erection and from cheaper procurement of material and equipment. A savings of 30 to 40 percent would appear to be valid.

CONCLUDING THOUGHTS

With regard to technologies or systems that may require special attention—catapults, side protection, compartmentation, or movement of weapons to the aircraft—MoD does not believe that any of these are sufficiently special to cause concern. MoD cites the availability of carrier-specific technologies, such as aircraft launch and recovery; industry efforts, primarily from the United States; and the possibility of customizing these technologies to meet its requirements for future aircraft carriers. The intellectual design ideas for other technologies can be studied and implemented without the in-depth knowledge required for such aircraft-specific technologies as launch and recovery.

In actuality, the changing shipbuilding environment in the UK and the evolving new design and production technologies may override any negative effects resulting from the size of the gap between design and production of aircraft-capable ships.

Appendix G

ISSUES OF AIRCRAFT CARRIER PRODUCTION IN FRANCE

This appendix documents the findings from a set of meetings with the French acquisition corps and members of their *Charles de Gaulle* aircraft carrier program at Paris and at the shipyard at Brest.

FRENCH DEFENSE ORGANIZATIONAL STRUCTURE

The French Ministry of Defense has two major branches: the General Chief of Staff, which encompasses all the French military forces, and the General Delegation for Armament (DGA), which develops, procures, and repairs the weapon systems for all the military forces. The DGA has operational directorates that align with major types of systems (land, naval, air) and functional directorates that provide specific capabilities across the weapon system acquisition community.[1]

One of the operational directorates of the DGA is the Directorate for Naval Construction (DCN), which is responsible for all naval ships, both submarine and surface. The DCN is the only branch of the DGA that has major industrial activities. It operates several shipyards throughout France that perform detailed design, construction, and maintenance functions.[2] The directorates for the air and land-based weapon systems either use the private sector (the Air Force) or state-owned (but not part of the DGA) facilities (the Army)[3] for their weapon system procurement and support.

---

[1]Since our meetings in France, the DGA has been reorganized. There are now three major operational directorates dealing mainly with program management (DSP for future and joint programs; DSA for management of conventional programs; and DPM for program management methods), plus several functional directorates (international relations, personnel, etc.) and three industrial entities (DCN for shipyards, DCE for test and evaluation, and SMA for aircraft maintenance).

[2]The directorate that has responsibility for aircraft has a few small depot-level maintenance activities in France.

[3]The state-owned Army production activities will soon shift to the private sector.

The DCN has three missions: (1) to provide the French Navy with combat ships and submarines required to fulfill the defense and public-service missions ordered by the French government; (2) to continuously maintain optimal performance levels of the ships and equipment in service; and (3) to provide its competence in the service of export programs and technical assistance to friendly foreign navies within the framework of government-defined political programs. To perform these missions, the DCN is divided into three subdirectorates: the Directorate for Naval Programs and International Cooperation, which provides the design, procurement, program management, and maintenance planning functions for the Navy; the Directorate for Industrial Activities, which manages the seven shipyards in France and provides a small staff for international trading activities; and the Directorate for Administrative Services, which provides staff administration and support functions.

The DCN is currently undergoing a major reorganization and downsizing in order to incorporate better business practices and to respond to the effects of reduced defense budgets. It is moving from a vertical structure, in which each service has divisions that provide the functional activities, to a horizontal structure organized around those functional activities (research and development, procurement, and maintenance). Downsizing within DCN has been difficult, whether at the Paris headquarters or at the various shipyards, since the civilian positions are civil service jobs governed by strict constraints on hiring and firing. Also, the shipyards are typically the major employers in their districts, and any reduction in force has a major impact on the regional economies.

The conceptual and initial design functions for submarines and surface ships are performed at DCN headquarters in Paris. The majority of the design efforts, including the final production design, are performed at the various shipyards. There are also design facilities at Toulon (DCN Engineering Center South) for combat systems and at Saint-Tropez for torpedoes. DCN Indret designs and builds the pressure vessels for nuclear reactors; Technicatome designs and builds the reactors.

DCN INDUSTRIAL FACILITIES

The DCN operates several industrial facilities throughout France.[4] The location and approximate personnel levels for these facilities are shown in Figure G.1.

[4]DCN also has a facility (approximately 400 personnel) at Papeete, Tahiti, to support the French naval vessels in the Pacific Ocean.

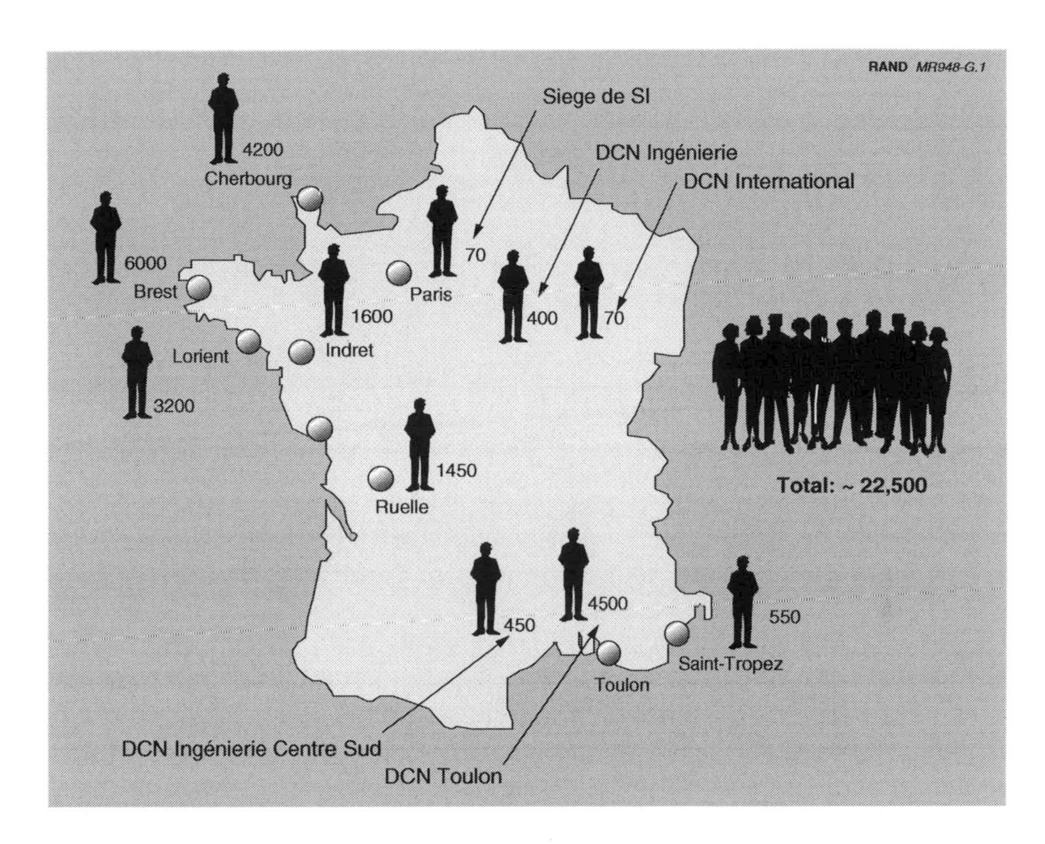

Figure G.1—DCN Industrial Activities in France

The French Navy is divided between an Atlantic Fleet homeported at Brest and the Mediterranean Fleet homeported at Toulon. These shipyards provide the logistics support and repair of the two fleets, as well as storing and maintaining their ammunition and missiles. DCN Brest is the primary shipyard for the construction of large surface ships, DCN Lorient constructs middle-tonnage ships, and DCN Cherbourg builds submarines. DCN Ruelle is the prime contractor for naval cybernetics, platform equipment, weapon-control systems, and missile launchers. DCN Saint-Tropez designs, produces, and tests torpedoes and countermeasure systems. DCN Toulon, in addition to supporting the ships of the French Mediterranean fleet, also has a center for combat system integration. DCN Indret designs, produces, and maintains naval propulsion systems, both conventional and nuclear (in conjunction with Technicatome).

Ship-procurement programs draw on the capabilities of all DCN facilities. Also, to balance workloads, construction may be shared across the various shipyards. For example, DCN Cherbourg and DCN Brest have constructed portions of frigates that were assembled and outfitted at DCN Lorient.

Currently, there are approximately 22,500 employees (down from approximately 26,600 in 1993) at DCN facilities (including Paris). These are primarily civil service employees; a small segment is uniformed personnel. The public, civil service nature of the shipyards makes it difficult for DCN to control their labor force, owing to tight restrictions on releasing civil service employees.

DCN BREST

DCN Brest is the largest public shipyard in France. In addition to constructing large surface ships, Brest is the home port for all French ships based in the Atlantic Ocean, including the French nuclear ballistic-missile submarine (SSBN) fleet. As such, it provides maintenance and repair functions for these ships, in addition to storing and maintaining their ammunition and missiles.

Brest has approximately 5,800 employees: 4,100 shipyard workers and 1,700 engineers (both ship design and integration/maintenance of electronic equipment) and management personnel. The shipyard is organized around three primary departments—Studies, Production, and Project Management—and its workload is divided almost equally between the maintenance of the surface fleet, the maintenance of the SSBNs, and new-ship construction. A small workload (about 10 percent of the total) is dedicated to missile and ammunition maintenance. These 5,800 civil service and uniformed personnel are augmented by both temporary hires (approximately 1,500, depending on budget availability) and subcontractors (approximately 500) who perform specific functions (e.g., sandblasting and painting) or are responsible for specific subsystems of a ship.

As the shipyard at Brest is downsized, excess DCN Brest employees are being offered positions with the French Navy that were formerly filled by conscripts. (Conscripts in the French Navy are being phased out.) These French Navy positions will be land-based. The ships constructed at Brest, along with the major reworks for SSBNs over the past 20 years, are listed in Figure G.2. Two large ships are currently under construction at Brest, the LPD *Siroco* and the CVN *Charles de Gaulle*. Brest is also building segments of Lafayette-class frigates that will be assembled in Lorient and that are intended for foreign military sales to Saudi Arabia and Taiwan.

The LPD *Siroco,* the second of the Foudre-class ships built at Brest, is 168 meters in length and displaces approximately 12,000 tons. It has provisions for 10 equipment-transportation barges, six heavy-duty tanks, and 23 light tanks, in addition to four helicopters of the Super Puma type. There are accommodations for almost 500 personnel. The first of its class, *Foudre* was commissioned in 1990 after a 6-year build period. *Siroco* is scheduled for commissioning

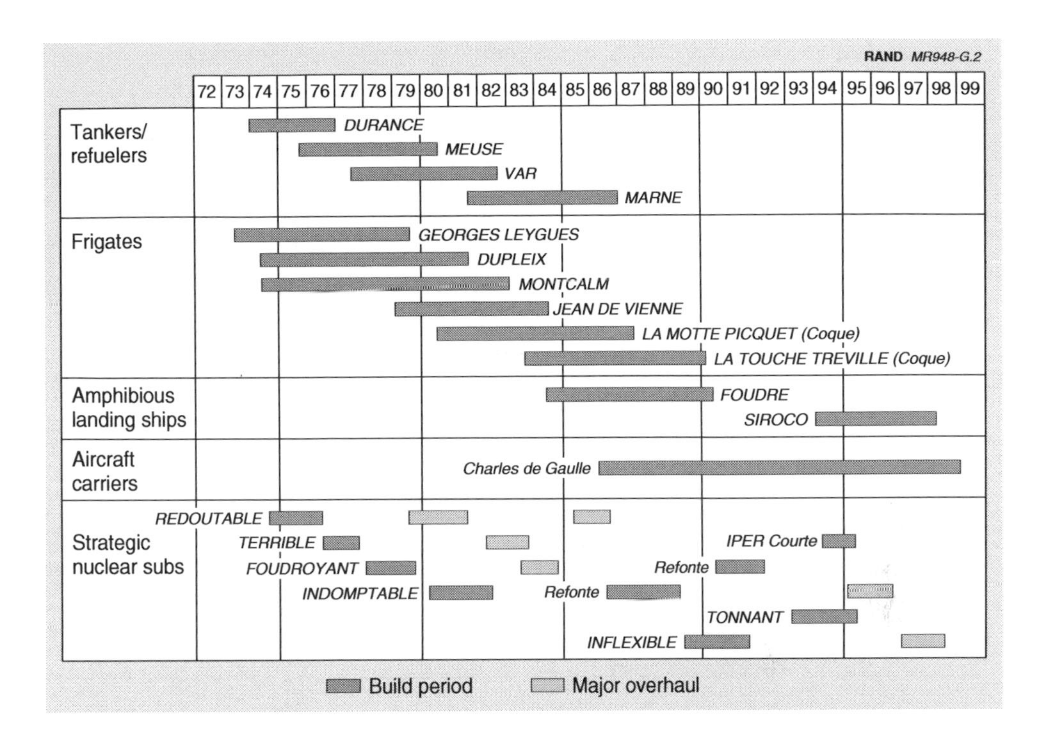

Figure G.2—Brest Ship Construction and Overhauls

in February 1998, after a 4-year construction period. In addition to the reduced construction time of *Siroco*, Brest officials state that the cost of *Siroco* is approximately 20 percent less than the cost of *Foudre*.

As mentioned, Brest also provides maintenance on the ships based there and is the only DCN shipyard that performs both new construction and maintenance. A large part of the maintenance workload is for the SSBN fleet, both for routine maintenance between patrols and for major overhaul and modernization.[5] At any one time, one of the SSBNs is in major-overhaul status. Each SSBN goes through a major overhaul and modernization approximately once every six years, and each major overhaul lasts 18 to 24 months.

THE *CHARLES DE GAULLE* PROGRAM

The French Navy currently operates two conventionally powered aircraft carriers, *Clemenceau* and *Foch*. These carriers are approximately 168 meters in

[5]The French Navy currently has five of the L'Inflexible class. The first of the new Le Triompant class was scheduled to become operational in 1996.

length and displace approximately 33,000 tons, fully loaded. The build period for these carriers is shown in Table G.1.

In 1980, the French Defense Council decided to build two nuclear-powered carriers to replace *Clemenceau* in 1996 and *Foch* some years later. The decision to build a nuclear-powered ship was motivated by the then-Soviet threat, concern about the availability and cost of diesel fuel, and the reduced overhead costs to France's submarine program because of the larger nuclear industrial base. The first of this class, named *Charles de Gaulle* (also referred to as *de Gaulle*), was ordered in February 1986; construction was started in November 1987. The keel was laid in April 1989, and the ship was launched in May 1994. Originally planned for delivery in 1996, budget problems have caused the program to stretch on, with a current planned commissioning date in 1999, at the earliest. The second carrier of the class has not yet been approved for construction, although it is recognized that *Foch* will go out of service early in the twenty-first century and that two carriers will be needed to provide continuous capability.[6]

De Gaulle will be approximately 261 meters long and have a displacement of approximately 40,000 tons.[7] It will accommodate 40 aircraft, including the new Rafale SU 0 class, the Super Etendard (to be replaced by the Rafale SU 2 in 2005), and E-2C Hawkeye early-warning aircraft, as well as several helicopters. There are two catapults, each capable of launching an aircraft every minute.[8] Propulsion is provided by two nuclear reactors of the same design as those used for the new-generation SSBNs.

Table G.1

Construction Periods for Clemenceau-Class Carriers

Ship	Builders	Laid Down	Launched	Commissioned
Clemenceau	DCN Brest	Sept 1955	21 Dec 1957	22 Nov 1961
Foch	Chantiers de l'Atlantique	Feb 1957	28 July 1960	15 July 1963

SOURCE: *Jane's All the World's Ships*, London: Sampson, Low, Marston and Co., 1993–1994.

[6]Whether the second ship will be nuclear-powered is still undecided.

[7]The size of *de Gaulle* was limited by the capacity of the dry docks at Brest, where she is being built, and at Toulon, where she will be homeported and maintained.

[8]The catapults are of American design but were built in France.

The ship's company was originally planned to be approximately 1,500 persons, but that number has been reduced to about 1,150 by functional analysis of the workloads and through automation.[9]

The initial design for *de Gaulle* was performed at the DCN facilities in Paris; Brest provided the detailed and final-production designs. Toulon has responsibility for the combat systems; Indret has design and production responsibility for the nuclear-propulsion system.

Funding problems have caused an approximate 4-year delay in the delivery of *de Gaulle*. Since the shipyard employees are civil servants, construction was never completely stopped. The funding problems resulted in delay of the services and capabilities provided by the temporary employees and subcontractors.

If they had had additional resources, the program managers would have explored and used more commercial standards and had more interaction with other governments and commercial firms. They are using fiber optics and commercial valves and pumps on board *de Gaulle*, and have the following turnkey systems:

- Fresh water (hot water, pressurized)
- Galley
- Laundry
- List-compensation system
- Commercial computers (Hewlett-Packard)
- Commercial radars for navigation.

FRENCH NAVAL NUCLEAR PROPULSION AND SAFETY

The industrial organization of French naval nuclear-propulsion production is shown in Figure G.3. Two ministries are involved, the Ministry of Defense (MoD) and the Ministry of Industry (MOI). MOI functions as does the U.S. Department of Energy and develops the technology. The STXN organization is equivalent to the U.S. Naval Nuclear Reactors organization and is responsible for hardware development, testing, and prototypes. Figure G.4 outlines the French safety organization.

[9]In addition to its company, the ship will host 550 aircrew and 50 flag staff; total accommodations are for 1,950, which includes the potential for 800 Marines.

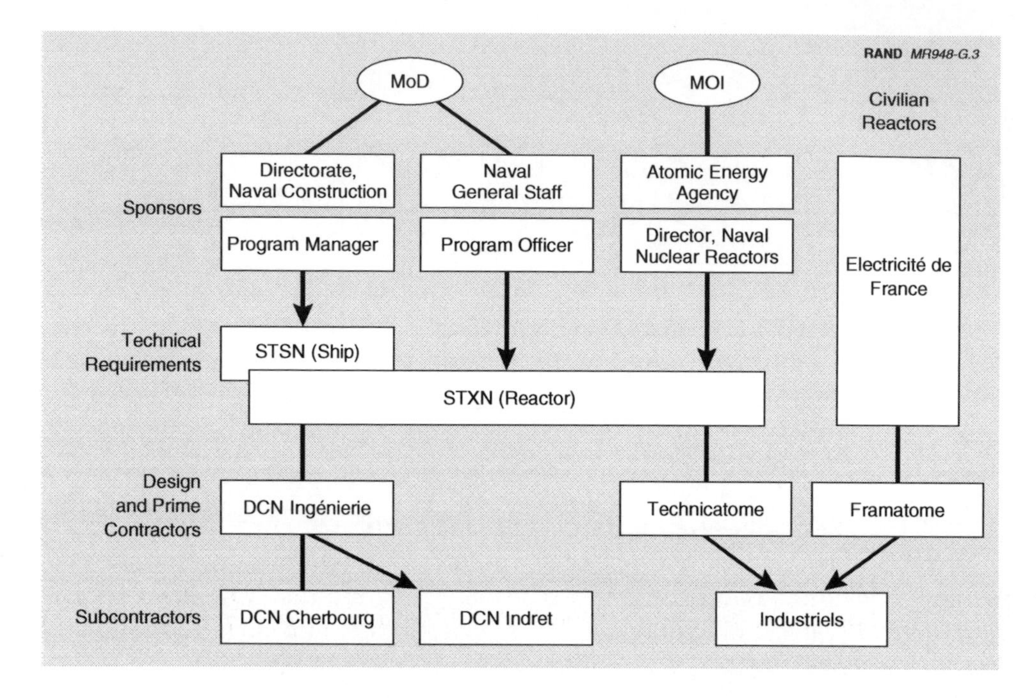

Figure G.3—French Naval Nuclear-Propulsion Organization

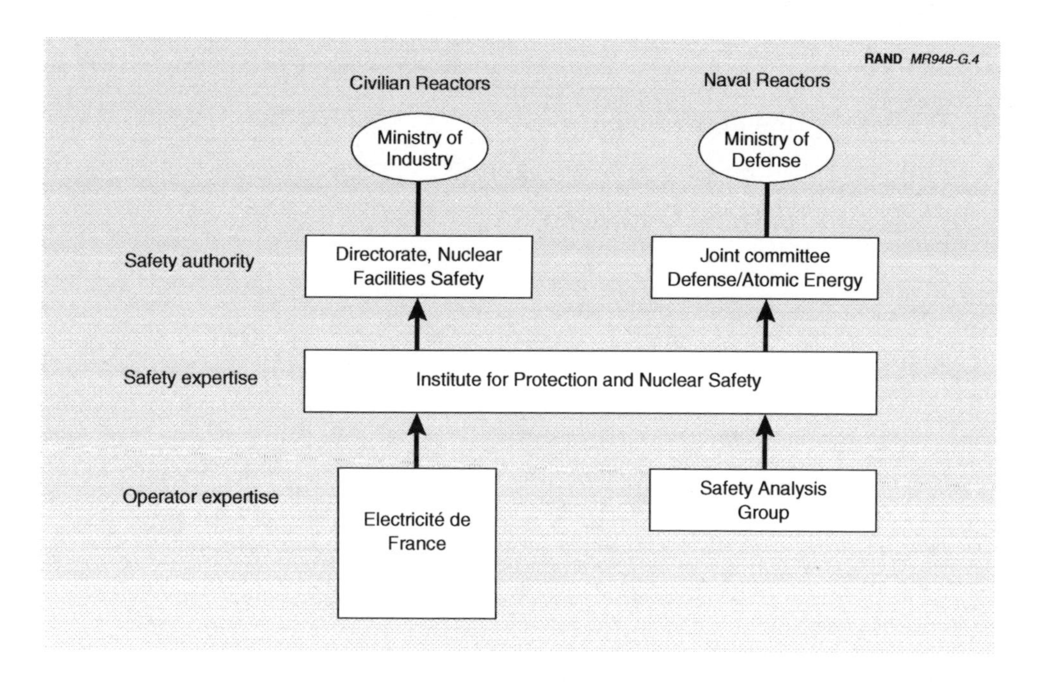

Figure G.4—French Nuclear Reactors Safety Control Organization

The Institute for Protection and Nuclear Safety must grant approval prior to start-up of either a commercial or nuclear reactor. In addition, this organization has authority to ask questions and to require reports and investigations of either military or commercial in-service reactors.

OBSERVATIONS

Our main reason for interacting with the DCN and officials of the *Charles de Gaulle* program was to understand any difficulties and problems associated with constructing a ship as large as an aircraft carrier after a production hiatus of almost two decades. Project personnel felt that the gap in carrier production did not cause any out-of-the-ordinary problems in the construction of *de Gaulle.* The shipyard had been active in building ships, although of a smaller size than a carrier, and these activities provided a sufficient foundation for construction of *de Gaulle.*

Social policies—e.g., maintaining fixed employment levels rather than workload—set the DCN's shipyard personnel levels. When funding is decreased, contractor activities are cut. For example, during one period of austere funding, the aft portion of the ship was sealed off and work continued in the forward sections. When problems did arise with systems unique to carrier construction, DCN turned to the United States for assistance. And for several systems, DCN contracted with U.S. firms that provide similar equipment for U.S. carriers.

The nuclear industry had been sustained by the submarine programs (in addition to a robust civilian nuclear-power industry), and it had taken advantage of the submarine reactor design and development efforts. The only subsystems that had caused problems were those peculiar to aircraft—primarily, the catapults. The French had to rely on American expertise in this area and made several trips to U.S. Navy facilities to gain knowledge and assistance.[10]

It is difficult to understand the economic impact of the long gap in the French carrier construction. No data were made available on the cost (or man-hours) of building *Charles de Gaulle;* however, the delays caused by lack of funding undoubtedly had a major effect on cost. One thing is certain: *de Gaulle has been built* after a long gap, suggesting that it is possible to stop building aircraft carriers for a long time and then reconstitute the capability. But this possibility assumes that the shipyard is active in the construction of large surface ships and maintains a nuclear-construction capability—and has access to an indus-

[10]This point should not be lost in any analysis of U.S. industrial capabilities: U.S. allies often rely on the United States for specific expertise that is fragile or expensive to maintain.

trial base that has been designing and manufacturing special equipment such as catapults.

In summary, French social policies, relative size of the ship (approximately one-third that of a Nimitz-class carrier), willingness to lengthen the production period, and availability of U.S. expertise and experience with carrier-unique systems obviated their need to worry about the availability of skilled labor.

What is not clear from the French experience is how such a production gap affects cost and the quality of the finished product.

Appendix H

CRUISE-SHIP PRODUCTION AT KVAERNER MASA HELSINKI NEW SHIPYARD

In addition to the discussion with builders of military ships, RAND investigated innovative commercial construction techniques that may lead to reduced carrier-construction costs. Of the commercial shipyards we could have studied, we focused on those building large cruise ships, because such ships are the most like aircraft carriers in size, electrical loads, and habitability (berthing, laundry, food service, waste handling and disposal, medical and dental support services, etc.).

In an industry experiencing overcapacity and generally low price levels, only a few commercial cruise-ship builders have succeeded. This appendix documents discussions with Kvaerner Masa in Helsinki.

ORGANIZATION AND PRODUCT

Kvaerner Masa-Yards is a Finnish shipbuilding company formerly owned by Wartsilla, a diesel-engine manufacturer, and acquired by Kvaerner A.S. in 1990. It builds cruise liners and passenger ferries, gas carriers, and ice breakers, as well as all types of special-technology vessels, such as special tankers, cable ships, research vessels, vessels for the offshore oil and gas industry, dredgers and cranes, and heavy-lift ships.[1] The company, which employs 4,900 people, has been a part of the Shipbuilding Division of the international industrial Kvaerner A.S. group since 1991. Kvaerner is headquartered in London. Other major divisions are

- the Kvaerner Masa-Yard Piikkiö Works, which is a factory producing prefabricated cabin and bathroom modules

[1]Kvaerner Masa-Yards and Kvaerner Masa Marine Inc., a subsidiary operating out of Annapolis, Maryland, are providing design and engineering for the 30,000-s.h.p. (shaft horsepower) Polar Icebreaker *Healy*, under construction at Avondale Industries Inc. for the U.S. Coast Guard.

- Kvaerner Masa-Azipod, which is responsible for the development, production, and sales of electric azimuthing Azipod propulsion drives[2]
- Kvaerner Masa-Yards Technology, which covers R&D, concept design and engineering services, and shipyard and welding technology, and includes the Arctic Technology Center (MARC) in Helsinki and the Welding Technology unit in Turku
- Kvaerner Masa Marine Inc., Vancouver, British Columbia, and its affiliate company Kvaerner Masa Marine Inc., Annapolis, Maryland, which are engaged in marine consulting engineering and marketing in North America.

Kvaerner Masa-Yards Inc. operates two new shipbuilding yards—Turku New Shipyard and Helsinki New Shipyard. Each yard is complete and independent: Each does its own design, engineering, and building, but the workload can be balanced between yards to ensure efficient workloading and profitability. Kvaerner Masa-Yards is in new-ship building only. Each had a repair capability but shed it because they are located out of the mainstream of ship traffic and because the Helsinki harbor is iced over during the winter months.

Kvaerner Masa-Yards' Helsinki New Shipyard is building the largest series of cruise ships ever ordered: eight ships for Carnival Cruise Lines, Inc. (see Table H.1). *Fantasy, Ecstasy,* and *Sensation* were delivered in 1990, 1991, and 1992. *Fascination* was delivered in 1994, and *Imagination* was delivered in 1995. *Inspiration* was delivered in 1996; the seventh, *Elation,* and the eighth, *Paradise,* were delivered in 1998. Each ship accommodates about 2,600 passengers, plus approximately 1,000 crew.

In addition, the cruise liner *Grandeur of the Seas* was delivered in 1996, and *Enchantment of the Seas* in 1997. These two were built for Royal Caribbean Cruises Ltd. Each accommodates about 2,400 passengers, plus about 800 crew, and has a gross tonnage of 74,000 tons and an overall length of 279.1 meters.[3]

---

[2]The Azipod propulsion unit, azimuthing through 360°, incorporates an electric AC motor located inside the propeller pod. The whole pod rotates, so the assembly produces vectored thrust, obviating the need for a rudder. The motor is driven by electric current generated by diesel engines. Eliminating the shaft results in a major design flexibility and space savings. A major breakthrough for the Azipod propulsion was the decision by Carnival Cruise Lines, Inc., to select Azipod propulsion for two 70,400 Fantasy-class cruise liners. Each of the cruise liners, due for delivery in 1998, will be fitted with two 14-megawatt Azipod units, which will result in useful space and weight savings on board, and will improve the ship's fuel efficiency.

[3]Between 1990 and 1996, Fincantieri, a rival builder of cruise ships, won orders for a total of 19 cruise vessels having an aggregate value of more than 10 trillion lire ($6.6 billion). When Walt Disney decided to enter the cruise business in spring 1995, it turned to Fincantieri to build its first ships, the 85,000-ton *Disney Magic* and *Disney Wonder,* each of which can carry 2,400 passengers. The Italian state shipbuilder has almost 40 percent of the total current world order-book for cruise ships, more than twice its nearest rival, Kvaerner Masa. Fincantieri has delivered the 101,000-gross-ton, $400M *Carnival Destiny,* the biggest cruise ship ever at the time, able to carry 3,300 holiday-

Table H.1

Specifications of Eight Carnival Cruise Ships Being Built by Kvaerner Masa

Length overall	260.6 meters
Maximum beam	36.0 meters
Freeboard	53.6 meters
Draft	7.75 meters
Speed	22 knots
Gross tonnage	70,367 tons

NOTE: Tonnage definitions vary for different types of ships. For passenger ships, the term is *gross tonnage*; for tankers and bulk cargo ships, the term is *deadweight tonnage*; and for warships, the term is *displacement tonnage*. These are defined as follows:

Gross tonnage. A measure of the total volume of enclosed spaces in the ship. The volume-to-tonnage conversion is 100 cu ft/ton.

Deadweight tonnage. A measure of the total volume of the ship dedicated to carrying cargo, converted to tons of seawater (35 cu ft/ton).

Displacement tonnage. The volume of water displaced by the hull beneath the waterline, converted to tons of seawater (35 cu ft/ton).

CORE COMPETENCIES AT KVAERNER MASA

Kvaerner Masa is one of the most innovative and profitable commercial shipyards in the world today, consistently returning a profit of 15 percent on net sales. Very competitive in the cruise-ship-construction niche as well as in other large, complex ship areas (liquid-natural-gas [LNG] carriers, icebreakers, cable ships), Kvaerner views itself as having four strengths:

- Project coordination
- Basic design of complex ships
- Hull fabrication
- Integration, final outfitting, and test.

Project Coordination

Each ship has a project manager and a deputy. The owner usually has representatives in the yard. One customer has a staff of 35 (similar to Supervisors of

makers, to Carnival. An even bigger vessel, the 109,000-gross-ton *Grand Princess*, was delivered to P&O's Princess Cruises in 1998. See Christopher Reynolds, "Cruiseopolis: The Humongous Grand Princess, Latest in the 'Biggest Ship' Sweepstakes, Is a Veritable Floating City," *Los Angeles Times*, June 21, 1998, p. L1.

Shipbuilding [SUPSHIPs] at U.S. shipyards); the other major owner has only one representative.

Corporate overhead functions are kept to a minimum. For example, administrative staff is a grand total of 5 (including one lawyer and one cost estimator); marketing has 10 people. All other people are in profit centers and are direct-charges.

Basic Design of Complex Ships

To build a new design, the first of a new class, Kvaerner would have a contract design specification of a few drawings and maybe 400 pages (compared with 2,000 drawings and many tomes for the United States). It would then take 3 years from beginning of design to delivery for a 70,000-ton cruise ship to be constructed, and 2 years and 3 months for the second.

The design office in Helsinki has 130 designers, 60 of whom are engineers from all disciplines. It has several architects working because of the emphasis on interior design. The architectural desires make the naval architects' job very difficult: Large open spaces in the hull disrupt deck continuity and deprive the ship of its traditional source of longitudinal strength. Many large windows on the sides further rob bending strength. Nevertheless, designers are very customer-oriented and develop design solutions to customer demands.

Hull Fabrication

Construction of large, complex cruise ships requires significant facilities, such as graving docks, piers, heavy-lift cranes, and covered work areas for steel fabrication and erection. The facilities are optimized for a throughput of 25,000 tons of steel per year.

The yard has, for several years, worked with the mill that supplies the steel, so that the steel now comes cut to size, bent to shape, end-prepped for welding, and blasted and coated. The specifications are sent to the yard by computer. The steel throughput is about 20–25 thousand tons a year (a small part of the mill's capacity). The steel-storage area in the yard is very small—maybe 100 plates of various sizes—because the yard demands just-in-time delivery on almost a daily basis. This approach allows Kvaerner to avoid making capital investments in facilities.

While visiting the yard, we saw several areas where steel-fabrication functions, automated in other yards, were being performed by a small group of workers. The production manager explained that the yard did not automate unless the process was a critical step in the time to complete the ship or if the yard could prove that automation would result in savings. If task x could be performed by 8 people and timing was not critical, the return on investment for automating (and reducing manpower by, say, 50 percent, to 4) would not be there. Thus, the decision not to automate.

Integration, Final Outfitting, and Test

The company is very aggressive about outsourcing. At the moment, about 50–55 percent of the completed-ship cost is spent on contracts to buy parts and services from other entities that are better (and cheaper) at building and providing them than the yard. The key is to have a very clean, well-defined relationship with subcontractors and to avoid specifying to the contractor how to build something—not a new idea, of course, but one that Kvaerner has developed and is using very effectively.

Perhaps the best example of this approach is Kvaerner's use of ready-to-install "floorless" modular cabins and bathroom units, which are manufactured by their Piikkiö Works, an independent profit center specializing in the manufacture of ready-to-install modular cabins and bathroom units for ships, offshore platforms, and hotels. The modular-cabin process and design are protected by patents. The cabins are delivered to the yard (Helsinki or Turku) by truck, three at a time.[4] They are completely finished when delivered, including furniture, bathroom fixtures, carpet, and bed linen. It takes 10 man-hours to install each cabin. Helsinki needs about 2,200 cabins per year, so the production rate is quite high.

Kvaerner is considering expanding its outsourcing to buy more turnkey components, such as galleys, laundries, gambling rooms, and bridges. In many respects, Kvaerner Masa is an integrator of other people's work. It has found that the less it does itself in the yard and the more it integrates, the more money is made and the more risk is reduced.

[4] Piikkiö Works employs about 200 people and is located near the town of Turku. About 50 percent of Piikkiö Works' production goes to competitors' yards.

ITALIAN CRUISE-SHIP PRODUCTION BY FINCANTIERI

This appendix documents discussions with the Italian shipbuilder Fincantieri. Fincantieri Cantieri Navali Italiani S.p.A. is the primary ship-construction company in Italy, building both commercial and naval ships, and is one of the world leaders in the design and construction of cruise ships. It can trace its lineage back almost 200 years and has constructed over 7,000 vessels of all types. The current company was first established in 1959 as a holding company for the shipbuilding sector within the IRI Group.[1] It was converted to an operating company in 1984 under its current name and currently operates two separate divisions, one dedicated to designing and building commercial ships and the other to designing and constructing military ships for the Italian Navy and for sale to other countries. A third division, specializing in diesel-engine design and construction, has recently spun off as a separate company.

The company is owned by the government but is operated as a commercial business (similar to a government-owned, commercially operated [GOCO] organization in the United States).[2] The company operates several shipyards throughout Italy, employing approximately 9,500 people, including 550 managers and design engineers at the Trieste headquarters. Although the workforce has been reduced in the past 10 years, production, especially in cruise-ship construction, has increased. Fincantieri currently has approximately 40 percent of the world cruise-ship-construction business (measured in number of beds) and 15 percent of the European commercial market.

Fincantieri's decision to reenter the passenger-ship business at the end of the 1980s, after a 25-year absence, proved just in time to catch a rising tide of orders for very large cruise ships accommodating thousands of passengers. Fincantieri has also been helped by the weak Italian lira, which has allowed it to undercut

[1]IRI stands for *Instituto per la Ricostruzione Industriale*, the largest Italian industrial conglomerate owned by the state.

[2]Although government-owned, the company is not subject to civil service personnel rules and policies. It uses commercial business practices to reduce or expand its labor force.

not only the Finns but other rivals in Germany and France. During 1990–1995, Fincantieri showed a positive return on investment, from 2.2 percent (in 1990) to over 10 percent (in 1995).

As with other European countries, Fincantieri receives a subsidy of approximately 8 percent. However, this subsidy is scheduled to be eliminated in 1998. Also, company leaders consider themselves somewhat disadvantaged in the commercial shipbuilding market, since one of their major competitors, Meyer Werft, receives a higher subsidy for performing some construction in East Germany (where higher subsidies are permitted in order to encourage economic development of former Soviet Bloc countries) and since financing is more difficult to obtain in Italy than in Northern European countries.

Fincantieri's Naval Shipbuilding Division has headquarters in Genoa and operates shipyards at Muggiano and Riva Trigoso, both on the Mediterranean coastline of northern Italy. It has constructed all types of ships for the Italian Navy, including the aircraft carrier *Garibaldi* and submarines (both at the Monfalcone shipyard),[3] plus destroyers, frigates, corvettes, and patrol vessels. Because of the reduction in naval shipbuilding programs, Muggiano and Riva Trigoso are currently building fast ferry boats for the commercial market in order to maintain workload in the yards.

While the Naval Shipbuilding Division has suffered from reduced defense spending in Italy, the Merchant Shipbuilding Division has grown in sales in the past decade as a result of the increased market in cruise ships. The commercial division operates six shipyards throughout Italy. Monfalcone and Venice-Marghera on Italy's northern Adriatic coast specialize in cruise-ship construction. Trieste and Ancona on the Adriatic coast, Palermo in Sicily, and Castellammare di Stabia on the Mediterranean coast build other types of commercial ships, including tankers and cargo vessels.

As mentioned, Fincantieri is recognized as one of the premier builders of cruise ships in the world. It builds ships for all the major cruise-ship operators and has recently won a contract from Disney for two 85,000-gross-ton[4] cruise ships. In November 1996, Monfalcone delivered the then largest cruise ship in the world (at 101,000 gross tons), *Carnival Destiny*. Monfalcone has since finished construction on and delivered *Grand Princess* (109,000 gross tons), which surpasses *Carnival Destiny* as the largest cruise ship in the world.

---

[3]Although once prominent in the construction of naval ships because of its large dry dock, Monfalcone now specializes in the construction of cruise ships.

[4]Gross tonnage is a measure of the enclosed volume of the ship, converted to tons at 100 cu ft/ ton. It is the standard measure used for cruise ships.

The commercial-shipbuilding practices at Fincantieri are similar to the practices of other commercial yards in Europe. The shipyard views itself as building the structural (hull, mechanical, and electrical) portions of the ship; the majority of the outfitting is accomplished on a turnkey basis by subcontractors.[5] Subcontractors and suppliers account for approximately 75 percent of the cost of the ship and concentrate on the hotel functions, such as cabins, galleys, and entertainment areas. Fincantieri works closely with thc subcontractors during the design phase, specifying the customer's needs and the physical arrangements of the basic hull form and utility services.

Fincantieri has flexible and adaptable arrangements with suppliers, trying to take advantage of changing economic environments and other opportunities as they arise. It tries to have several suppliers for a given commodity, thereby fostering competition and, hopefully, lower prices. But it strives to maintain long-term relationships with suppliers, realizing that such relationships help to avoid unknowns and typically provide quality products at a fair price. It tries to purchase materials, such as pipes, and subcontractor items, such as cabins, in large quantities to again obtain lower prices. For example, the Monfalcone shipyard has relationships with four steel suppliers that provide steel plates cut to size.

Fincantieri estimates that the first of a class for the current-size cruise ships requires approximately 2 million man-hours of Fincantieri employee labor plus an additional 60 to 70 percent for subcontractors—a figure Fincantieri expects to decrease by about one-third for the second ship in a class. Construction time for current cruise ships is approximately 18 months from contract award to delivery. The approximate cost is $15 per kilogram, based on displacement tons.[6]

Monfalcone is the largest Fincantieri shipyard in dry-dock size (350 meters long, 56 meters wide, handling ships up to 300,000 deadweight tons), facilities, and employees (approximately 2,000). In fact, we were told that Monfalcone is the largest shipyard in Europe. The steel-fabrication facility is modern, clean, and well-lighted and well-ventilated. A modern automated fabrication system (flexible automated steel prefabrication [FASP]) uses robotic articulated single arms to weld stiffeners to the steel plate. The yard is in the process of installing a laser welding capability and a line heat-shaping facility. Although the yard uses modular construction (with an 800-ton gantry crane to place the modules into the dock), it does not pre-outfit modules to the degree that is common in other European and American shipyards. For example, electric cables are

[5]Fincantieri says that a system in which it contracts out 70 percent of the workload to a network of local subcontractors enables it to deliver ships more punctually than can its rivals.

[6]As an example, using the $15-per-kilogram rule of thumb, we see that *Grand Princess*, which displaces approximately 28,000 tons, would have cost $420 million. This cost metric is comparable to that of other European commercial shipbuilders.

installed after launch instead of during pre-outfitting of the modules (although cable trays are installed during pre-outfitting). Shipyard managers stated this was done since the splicing of cables was not permitted.

Other construction in current ships is stainless low-pressure piping, joined with pressed or swaged joints, which is used throughout the ships; circular, low-cross-section ventilation ducting with double-walled construction and foam insulation to reduce noise; and composite materials, which are being used experimentally in the Disney cruise ships, in one of the stacks that houses public spaces. Ship designers and builders also place great emphasis on the internal arrangements for the handling and moving of passenger luggage and ship supplies. Innovative methods make it possible to move 3,000 passengers and 9,000 pieces of luggage off and on the ship in a matter of hours.

NIMITZ-CLASS AIRCRAFT CARRIER OPERATING AND SUPPORT COSTS

This appendix presents charts (Figures J.1 through J.20) showing Operating and Support (O&S) costs for Nimitz-class aircraft carriers. The data are taken from the Visibility and Management of Operating and Support Costs (VAMOSC) database. All costs shown here are in FY98 dollars. The VAMOSC database available at the time of this study covered FY 1978 through FY 1995, a time window that covers different portions of the lives of the individual Nimitz-class ships. Table J.1 shows the correspondence between fiscal year and ship age, by hull number. There are two sets of charts, one for top-line VAMOSC cost categories broken down by each of four costs for all ships and by all four costs for each ship, and one set for the four depot maintenance costs, broken down in the same way as the first set.

The first set of four charts (Figures J.1 through J.4) presents the four top-line VAMOSC cost categories for all the Nimitz-class ships: Annual Direct Unit Costs, Annual Direct Intermediate Maintenance Costs, Annual Direct Depot Maintenance Costs, and Annual Indirect O&S Costs. The next six charts (Figures J.5 through J.10) show the above four cost categories for each Nimitz-class ship and are plotted against the same y-axis scale to facilitate comparisons across ships. For all charts, the x-axis covers ship life from 1 to 20 years.

Table J.1

Ages of Nimitz-Class Ships for Period Covered by VAMOSC Database

Hull No.	Name	Age in FY78	Age in FY95
CVN 68	*Nimitz*	3	20
CVN 69	*Eisenhower*	1	18
CVN 70	*Carl Vinson*		13
CVN 71	*Theodore Roosevelt*		8
CVN 72	*Abraham Lincoln*		5
CVN 73	*George Washington*		3

The third set of charts (Figures J.11 through J.14) shows the four categories of depot maintenance costs: Annual Scheduled Ship Overhaul Costs, Annual Non-Scheduled Ship Repair Costs, Annual Fleet Modernization Costs, and Annual Other Depot Costs. The term "annual" in these categories is particularly important, because the availabilities referred to in the main text of this report frequently spread over two and sometimes more fiscal years. Hence, the costs shown in VAMOSC for any given year may not equal the cost of a given availability. Again, these four charts are followed by six charts (Figures J.15 through J.20), broken down by each ship for all four costs, against the same *y*-axis scale for comparison.

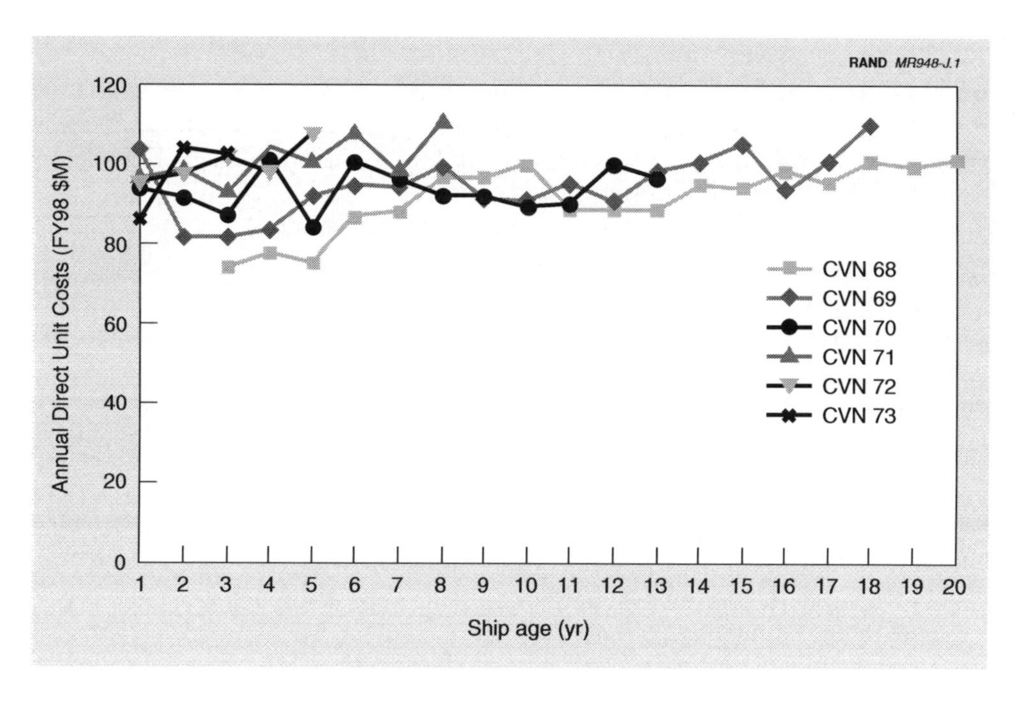

Figure J.1—Annual Direct Unit Costs for Each Nimitz-Class Ship, by Hull Number by Age

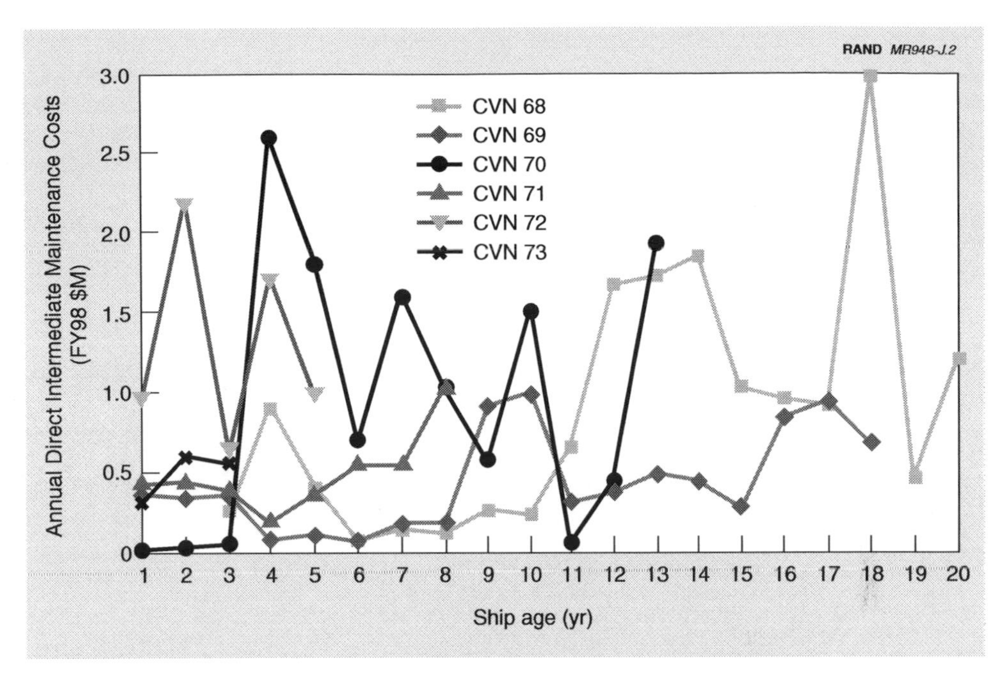

Figure J.2—Annual Direct Intermediate Maintenance Costs for Each Nimitz-Class Ship, by Hull Number by Age

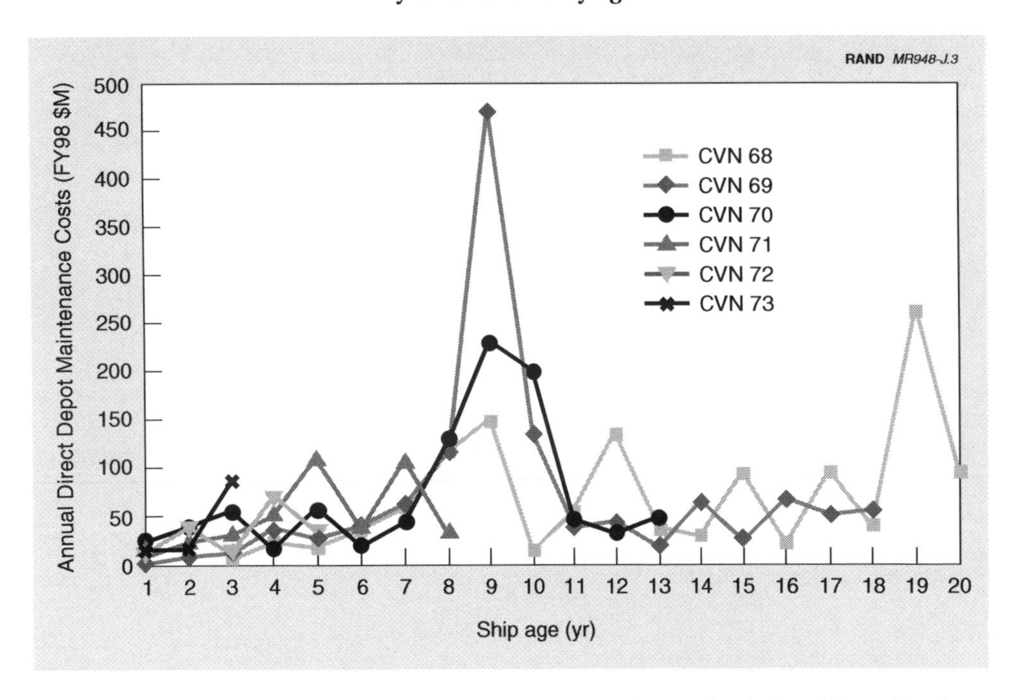

Figure J.3—Annual Direct Depot Maintenance Costs for Each Nimitz-Class Ship, by Hull Number by Age

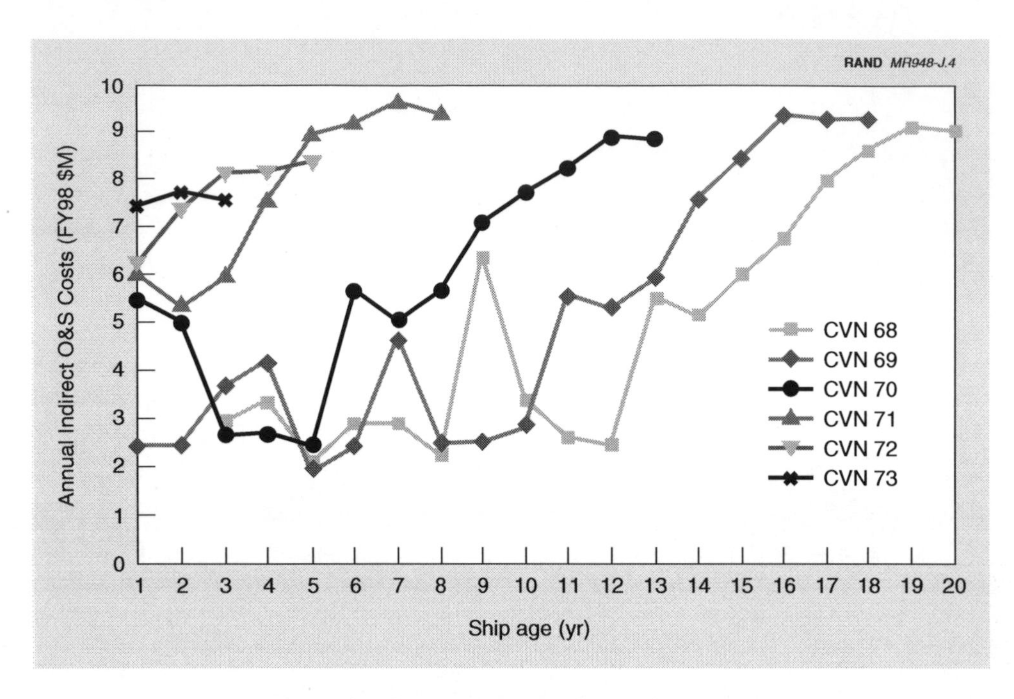

Figure J.4—Annual Indirect O&S Costs for Each Nimitz-Class Ship, by Hull Number by Age

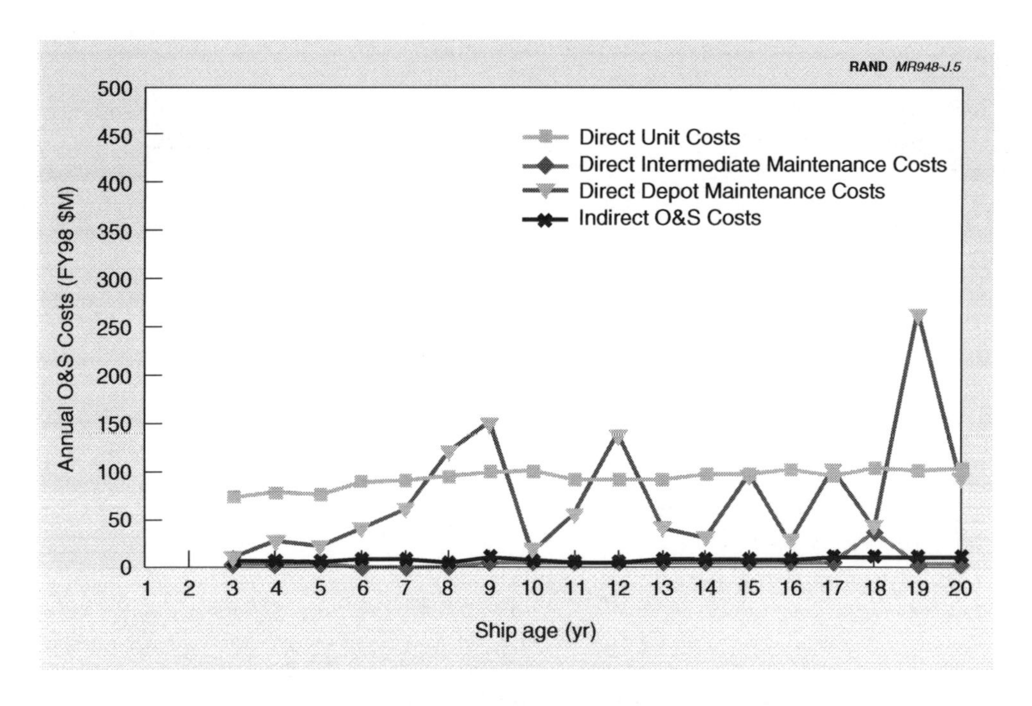

Figure J.5—Annual O&S Costs for CVN 68, by Cost Category by Age

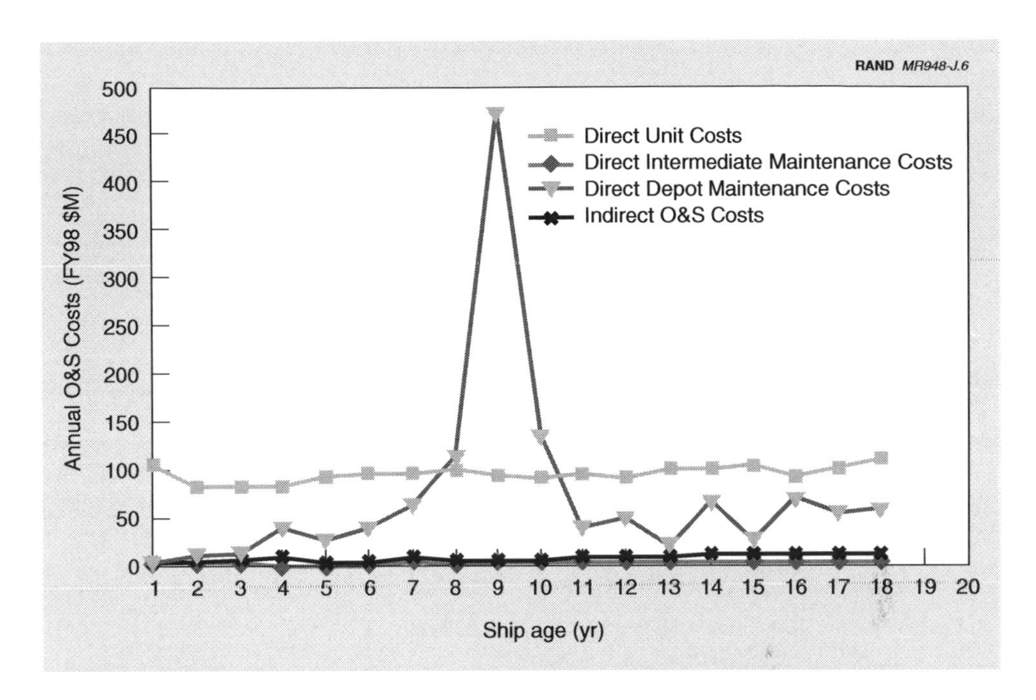

Figure J.6—Annual O&S Costs for CVN 69, by Cost Category by Age

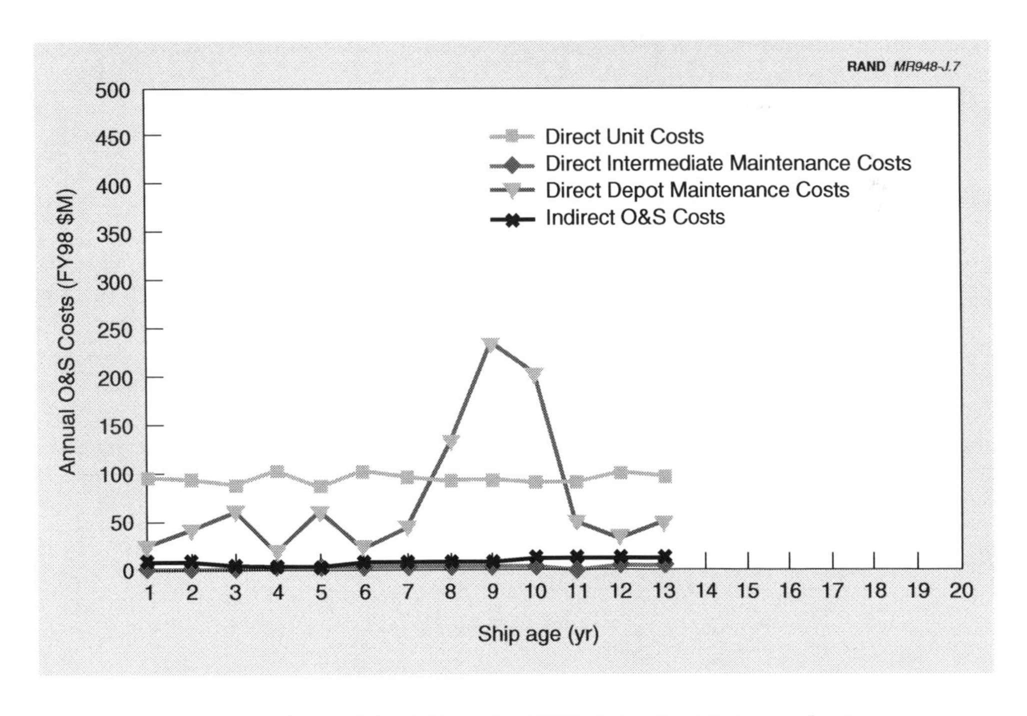

Figure J.7—Annual O&S Costs for CVN 70, by Cost Category by Age

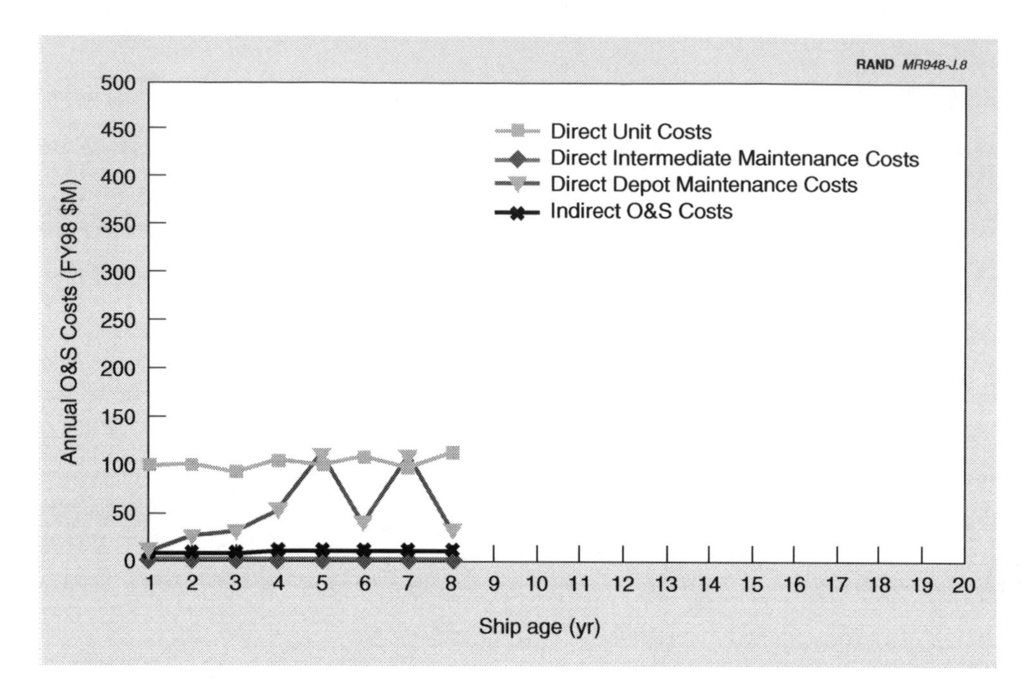

Figure J.8—Annual O&S Costs for CVN 71, by Cost Category by Age

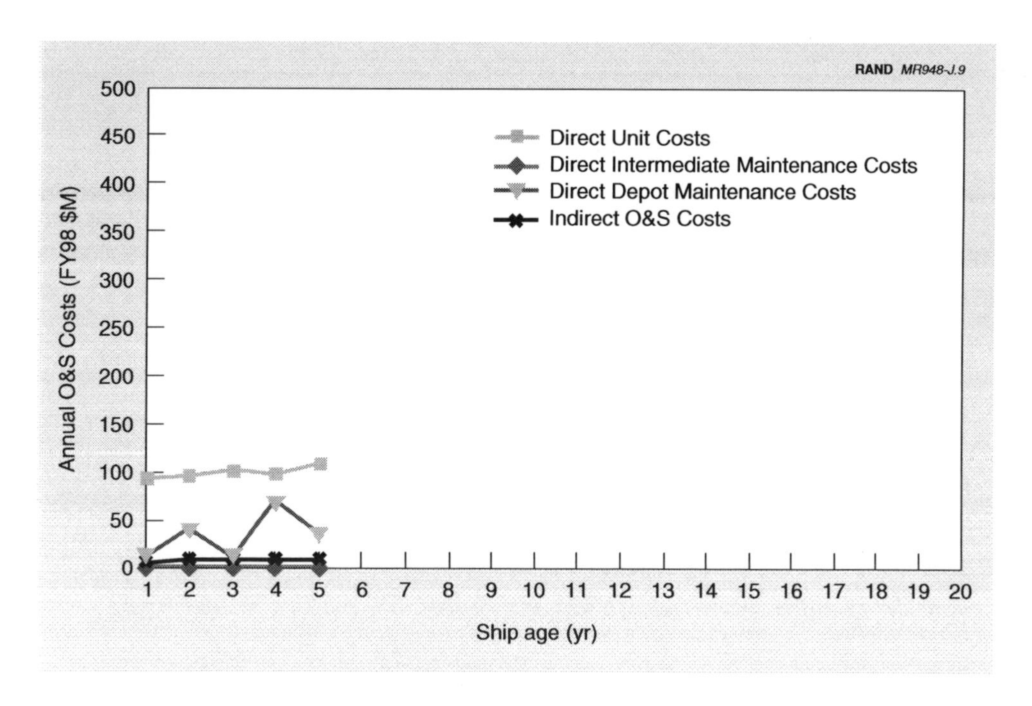

Figure J.9—Annual O&S Costs for CVN 72, by Cost Category by Age

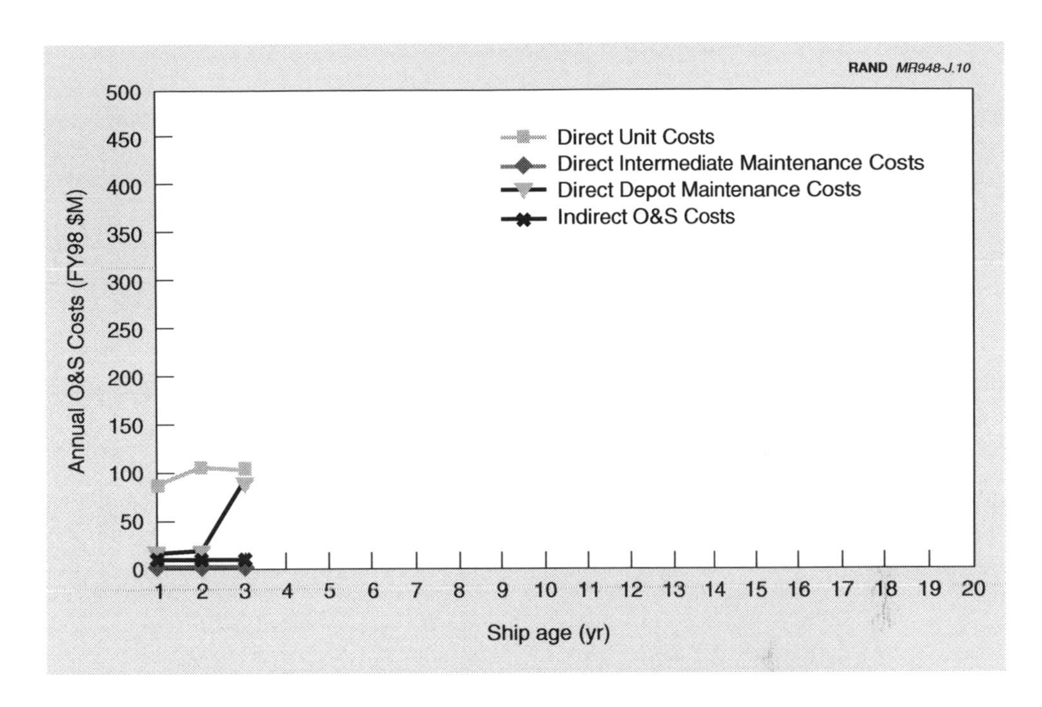

Figure J.10—Annual O&S Costs for CVN 73, by Cost Category by Age

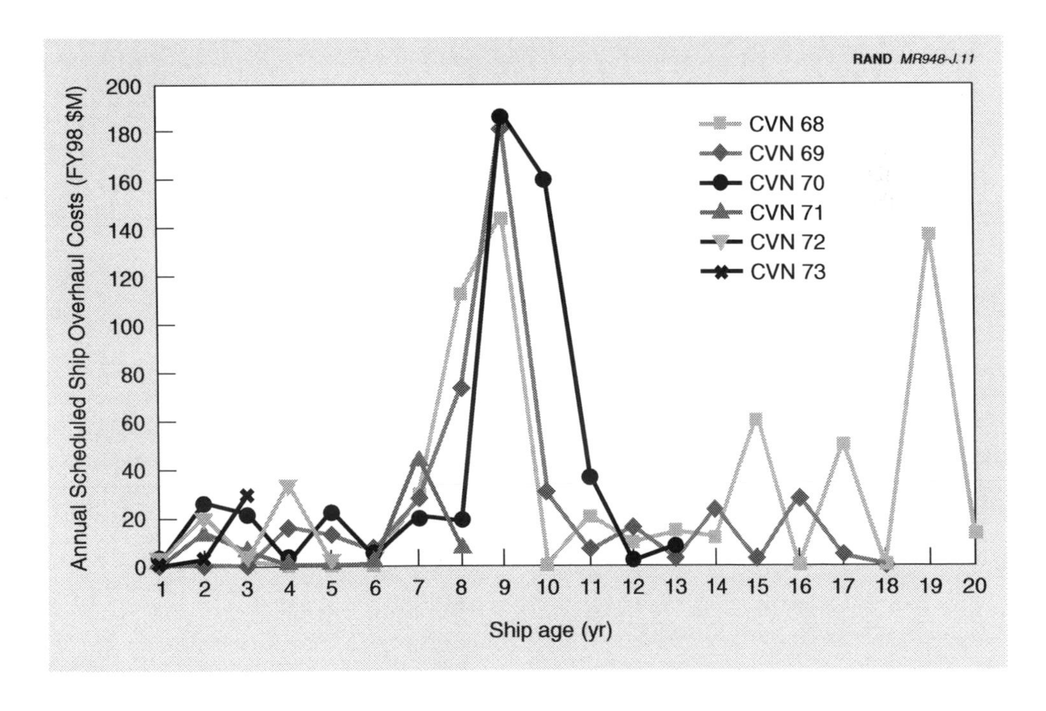

Figure J.11—Annual Scheduled Ship Overhaul Costs, by Hull Number by Age

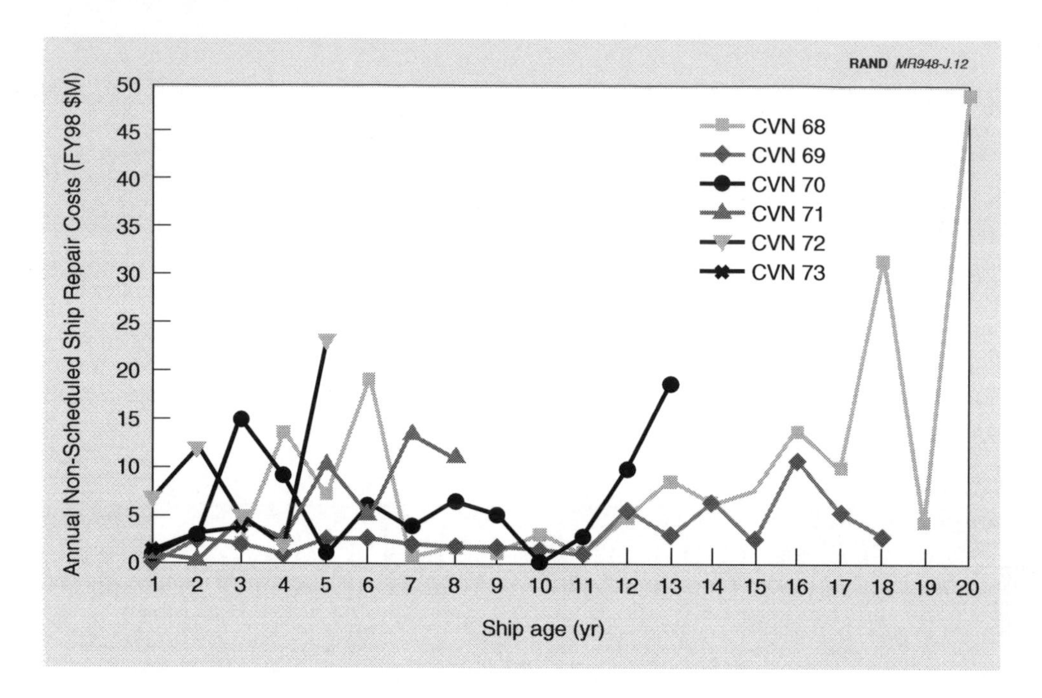

Figure J.12—Annual Non-Scheduled Ship Repair Costs, by Hull Number by Age

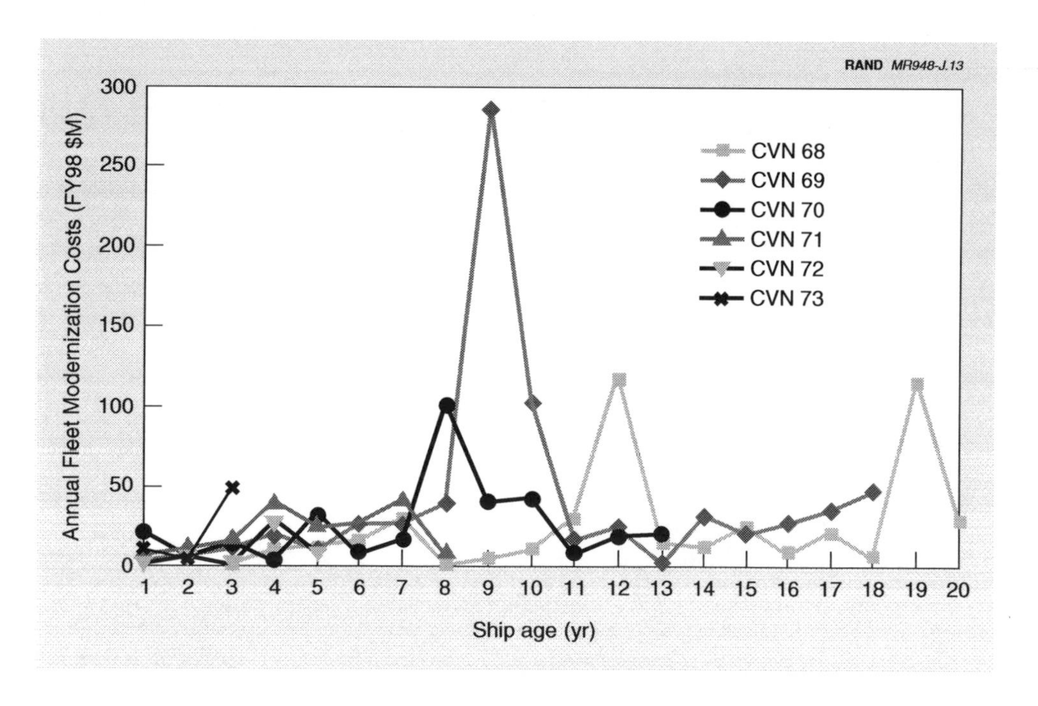

Figure J.13—Annual Fleet Modernization Costs, by Hull Number by Age

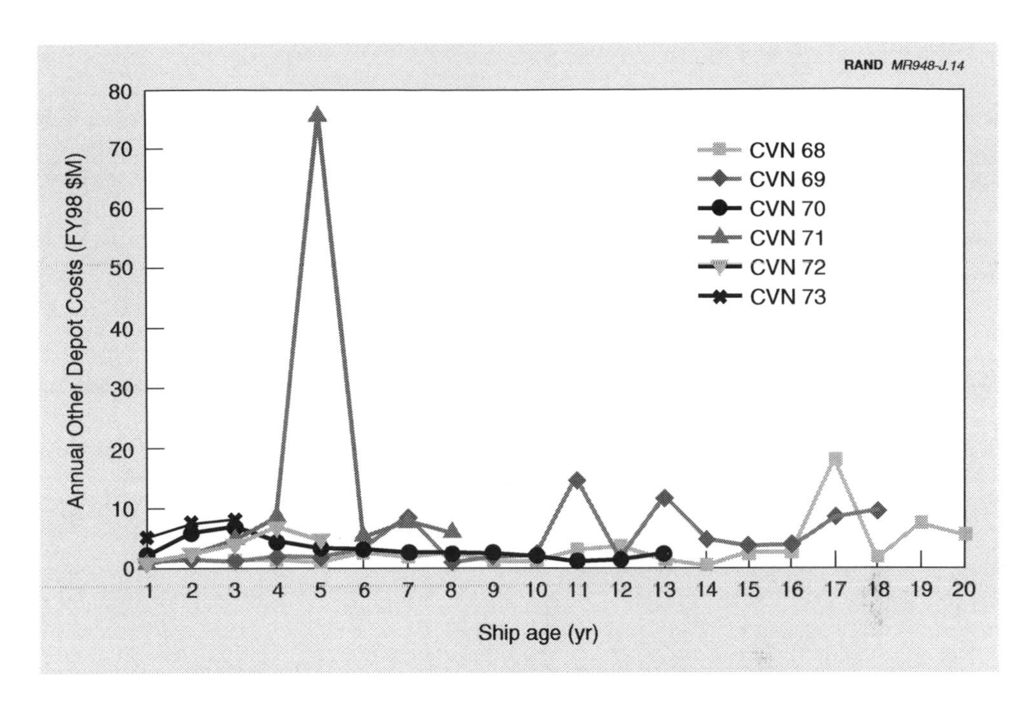

Figure J.14—Annual Other Depot Costs, by Hull Number by Age

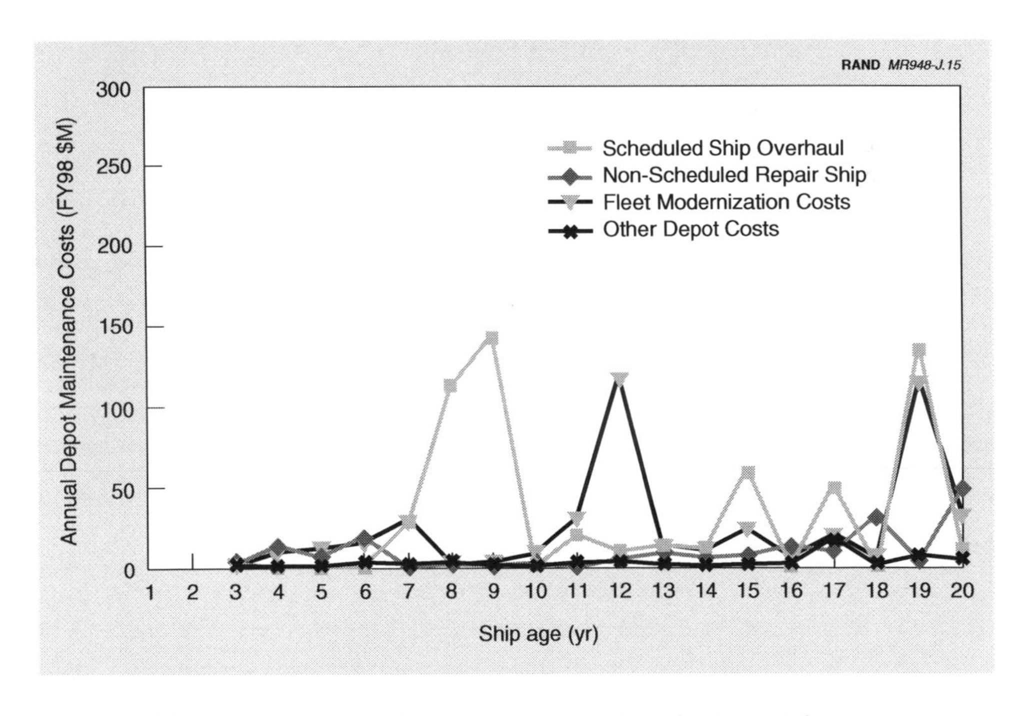

Figure J.15—Annual Depot Maintenance Costs for CVN 68, by Age

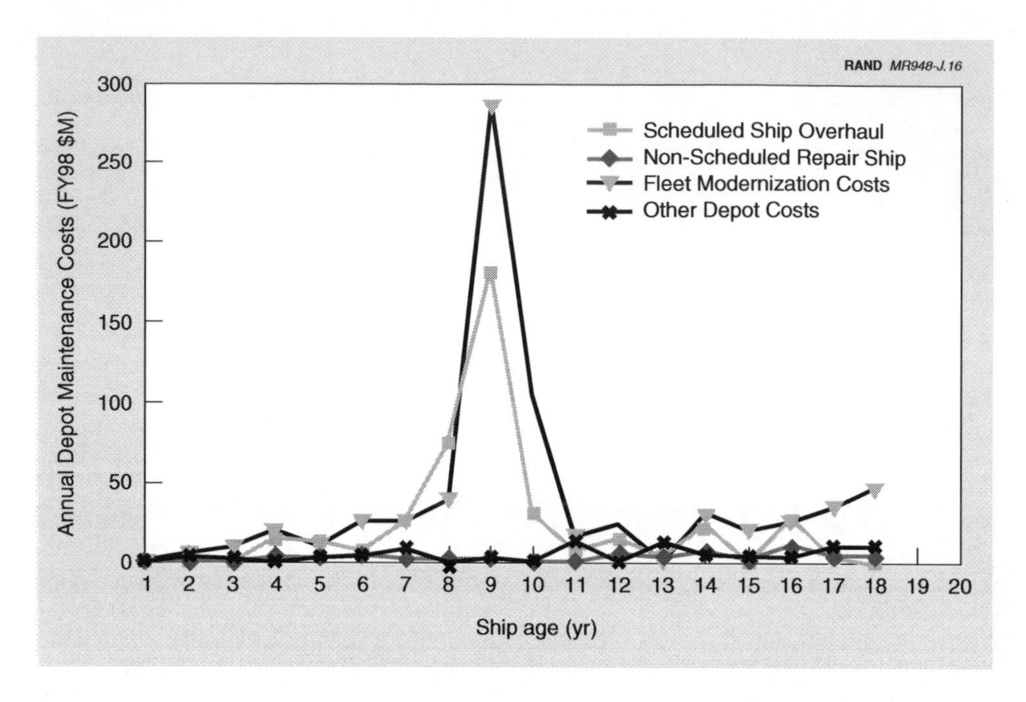

Figure J.16—Annual Depot Maintenance Costs for CVN 69, by Age

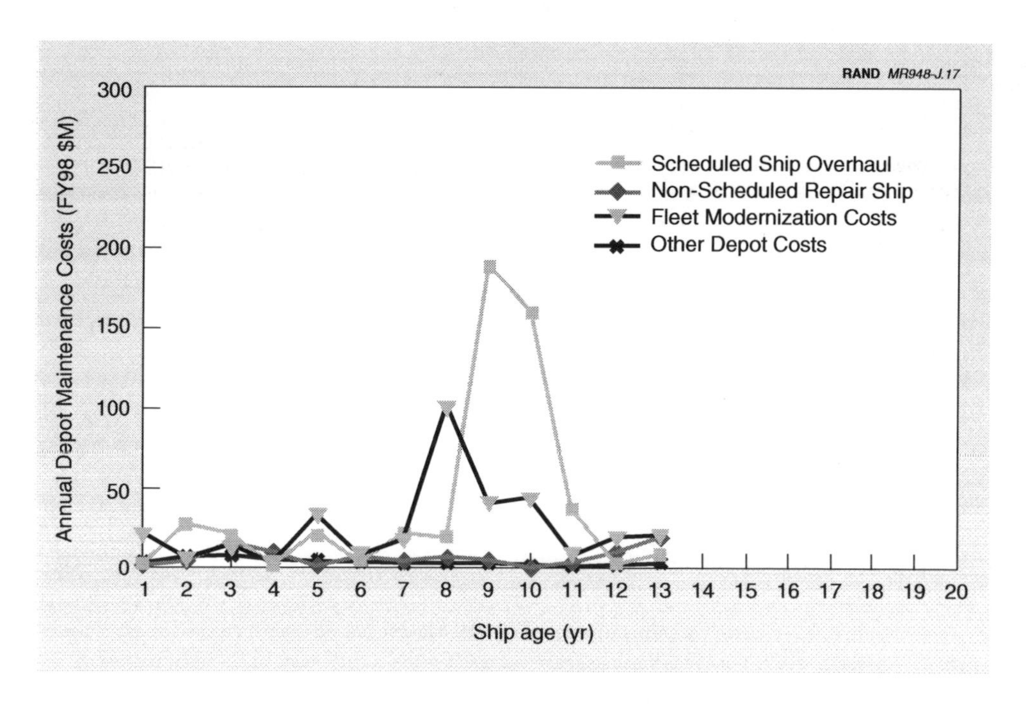

Figure J.17—Annual Depot Maintenance Costs for CVN 70, by Age

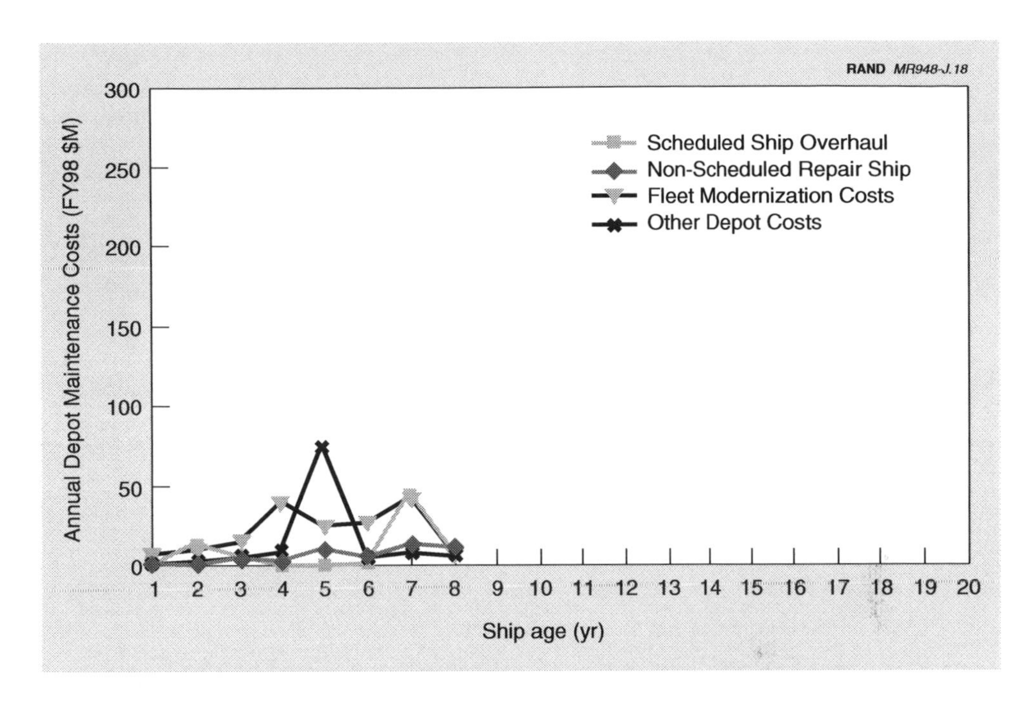

Figure J.18—Annual Depot Maintenance Costs for CVN 71, by Age

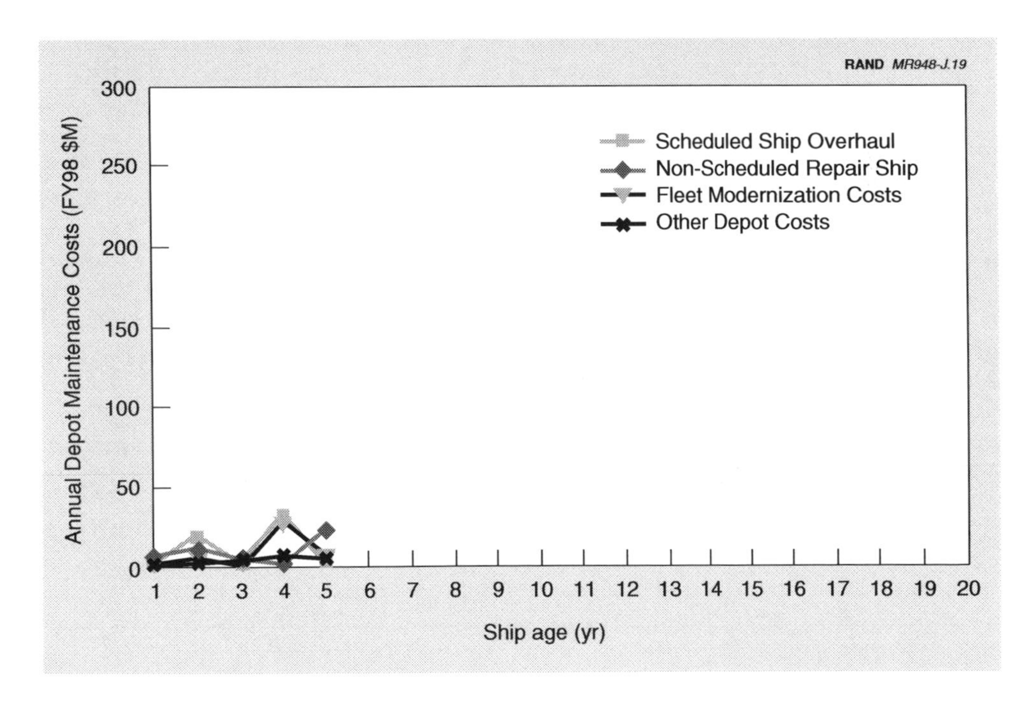

Figure J.19—Annual Depot Maintenance Costs for CVN 72, by Age

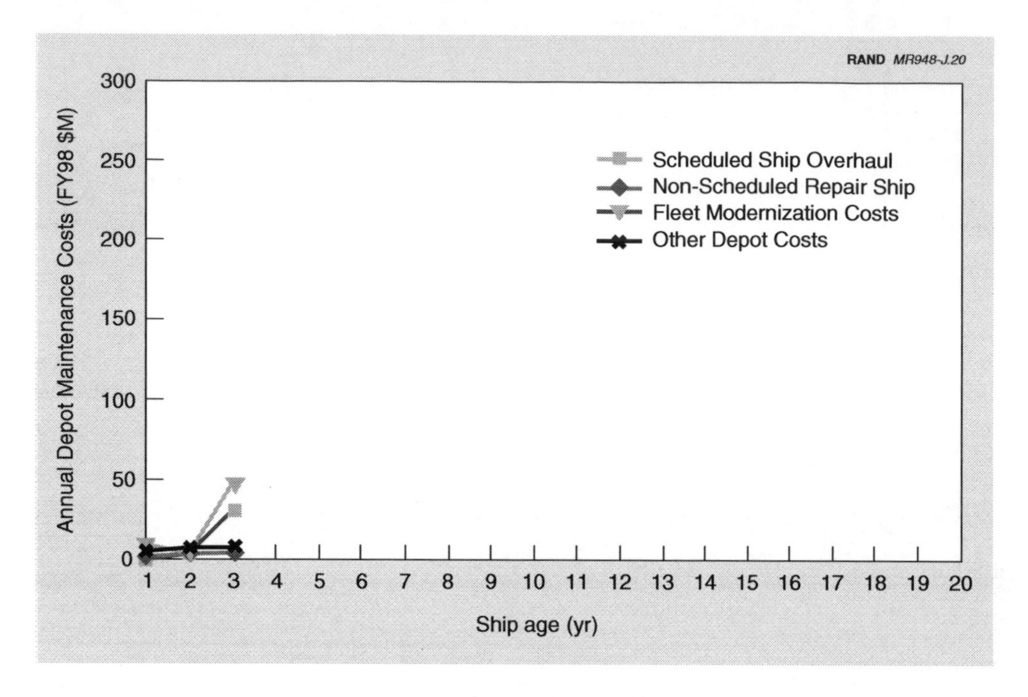

Figure J.20—Annual Depot Maintenance Costs for CVN 73, by Age

BIBLIOGRAPHY

Barnett, Thomas P.M., and Linda D. Lancaster, *Answering the 9-1-1 Call: U.S. Military and Naval Crisis Response Activity, 1977–91*, Alexandria, Va.: Center for Naval Analyses, ADB173802, August 1992.

Birkler, John, et al., *Preliminary Analysis of Industrial-Base Issues and Implications for Future Bomber Design and Production,* Santa Monica, Calif.: RAND, MR-628.0-AF, October 1995.

Birkler, John, et al., *Reconstituting a Production Capability: Past Experience, Restart Criteria, and Suggested Policies,* Santa Monica, Calif.: RAND, MR-273-ACQ, 1993.

Birkler, John, et al., *The U.S. Submarine Production Base: An Analysis of Cost, Schedule, and Risk for Selected Force Structures,* Santa Monica, Calif.: RAND, MR-456-OSD, 1994a.

Birkler, John, et al., *The U.S. Submarine Production Base: An Analysis of Cost, Schedule, and Risk for Selected Force Structures: Executive Summary,* Santa Monica, Calif.: RAND, MR-456/1-OSD, 1994b.

Chesneau, Roger, *Aircraft Carriers of the World, 1914 to the Present: An Illustrated Encyclopedia,* Annapolis, Md.: Naval Institute Press, 1984.

Davis, Jacquelyn K., *CVX: A Smart Carrier for the New Era,* Washington, D.C.: Brassey's, 1998.

Friedman, Norman, *U.S. Aircraft Carriers: An Illustrated Design History,* Annapolis, Md.: Naval Institute Press, 1983.

Jane's Fighting Ships , London: Jane's Information Group, various years.

Jane's Fighting Ships, London: Sampson Low, Marston and Co., 1996.

Lautenschlager, Karl, *Technology and the Evolution of Naval Warfare 1851–2001,* Washington, D.C.: National Academy Press, 1984.

Leopold, Reuven, *Sea-Based Aviation and the Next U.S. Aircraft Carrier Design: The CVX*, Cambridge, Mass.: Massachusetts Institute of Technology, Center for International Studies, MIT Security Studies Program, January 1998.

Naval Sea Systems Command, Naval Propulsion Directorate, *March 3 Report on Preservation of the U.S. Nuclear Submarine Capability*, Supplement to Naval Nuclear Industrial Base report of November 10, 1992; updated through interview with Naval Nuclear Propulsion Directorate.

Office of Management and Budget, *Memorandum for Heads of Executive Departments and Establishments (Subject: Guidelines and Discount Rates for Benefit-Cost Analysis of Federal Programs)*, Circular A-94 (revised), October 23, 1992, Appendix C (revised February 1997).

O'Rourke, Ronald, *Navy Major Shipbuilding Programs and Shipbuilders: Issues and Options for Congress*, Washington, D.C.: Congressional Research Service, September 24, 1996.

Perry, Robert, et al., *Development and Commercialization of the Light Water Reactor, 1946–1976*, Santa Monica, Calif.: RAND, R-2180-NSF, June 1977.

Planning, Engineering, Repairs, Alterations—Aircraft Carriers (PERA-CV), *Incremental Maintenance Program for CVN-68 Class Aircraft Carriers*, Bremerton, Wash., January 1, 1997.

Reynolds, Christopher, "Cruiseopolis: The Humongous Grand Princess, Latest in the 'Biggest Ship' Sweepstakes, Is a Veritable Floating City," *Los Angeles Times*, June 21, 1998, p. L1.

Shalikashvili, John, *National Military Strategy of the United States*, Washington, D.C.: Joint Chiefs of Staff, 1995.

Siegel, Adam B., *The Use of Naval Forces in the Post-War Era: U.S. Navy and U.S. Marine Corps Crisis Response Activity, 1946–1990*, Alexandria, Va.: Center for Naval Analyses, CRM-90-246, February 1991.

U.S. Naval Center for Cost Analysis, *Navy Visibility and Management of Operating and Support Cost (Navy VAMOSC): Data Reference Manual for Individual Ships Report*, Arlington, Va., April 30, 1997.

Vick, Alan, David T. Orletsky, Abram N. Shulsky, and John Stillion, *Preparing the U.S. Air Force for Military Operations Other Than War*, Santa Monica, Calif.: RAND, MR-842-AF, 1997.

The World Aircraft Carrier Lists, available online at http://www.uss-salem.org/navhist/carriers/.